Chemistry Research and Applications

MOLYBDENUM AND TUNGSTEN COFACTOR MODEL CHEMISTRY

CHEMISTRY RESEARCH AND APPLICATIONS

Handbook on Mass Spectrometry: Instrumentation, Data and Analysis, and Applications
J. K. Lang (Editor)
2009. 978-1-60741-580-0

Applied Electrochemistry
Vijay G. Singh (Editor)
2009. 978-1-60876-208-8

Handbook of Inorganic Chemistry Research
Desiree A. Morrison (Editor)
2010. 978-1-61668-010-7

Solid State Electrochemistry
Thomas G. Willard (Editor)
2010. 978-1-60876-429-7

Mathematical Chemistry
W. I. Hong (Editor)
2010. 978-1-60876-894-3

Physical Organic Chemistry: New Developments
Karl T. Burley (Editor)
2010. 978-1-61668-435-8

Chemical Sensors: Properties, Performance and Applications
Ronald V. Harrison (Editor)
2010. 978-1-60741-897-9

Macrocyclic Chemistry: New Research Developments
Dániel W. Fitzpatrick and Henry J. Ulrich (Editors)
2010. 978-1-60876-896-7

Rock Chemistry
Basilio Macías and Fidel Guajardo (Editors)
2010. 978-1-60876-563-8

Electrolysis: Theory, Types and Applications
Shing Kuai and Ji Meng (Editors)
2010. 978-1-60876-619-2

Energetic Materials: Chemistry, Hazards and Environmental Aspects
Jake R. Howell and Timothy E. Fletcher (Editors)
2010. 978-1-60876-267-5

Chemical Crystallography
Bryan L. Connelly (Editor)
2010. 978-1-60876-281-1

Heterocyclic Compounds: Synthesis, Properties and Applications
Kristian Nylund and Peder Johansson (Editors)
2010. 978-1-60876-368-9

Influence of the Solvents on Some Radical Reactions
Gennady E. Zaikov, Roman G. Makitra,
Galina G. Midyana and Lyubov.I Bazylyak (Editors)
2010. 978-1-60876-635-2

Dzhemilev Reaction in Organic and Organometallic Synthesis
Vladimir A.D'yakonov (Author)
2010. 978-1-60876-683-3

Analytical Chemistry of Cadmium:
Sample Pre-Treatment and Determination Methods
Antonio Moreda-Piñeiro and Jorge Moreda-Piñeiro (Authors)
2010. 978-1-60876-808-0

Binary Aqueous and CO2 Containing
Mixtures and the Krichevskii Parameter
Aziz I. Abdulagatov, Ilmutdin M. Abdulagatov,
Gennadii V. Stepanov (Authors)
2010. 978-1-60876-990-2

Advances in Adsorption Technology
Bidyut Baran Saha and Kim Choon Ng (Editors)
2010. 978-1-60876-833-2

Electrochemical Oxidation and Corrosion of Metals
Elena P. Grishina and Andrew V. Noskov (Authors)
2010. 978-1-61668-329-0

Modification and Preparation of Membrane in Supercritical Carbon Dioxide
Guang-Ming Qiu, Rui Tian, Yang Qiu and You-Yi Xu (Authors)
2010. 978-1-60876-905-6

Thermostable Polycyanurates: Synthesis, Modification, Structure and Properties
Alexander Fainleib (Editor)
2010. 978-1-60876-907-0

Combustion Synthesis of Advanced Materials
B. B. Khina (Author)
2010. 978-1-60876-977-3

Structure and Properties of Particulate-Filled Polymer Composites: The Fractal Analysis
G. V. Kozlov, Y. G. Yanovskii and G. E. Zaikov (Authors)
2010. 978-1-60876-999-5

Information Origins of the Chemical Bond
Roman F. Nalewajski (Author)
2010. 978-1-61668-305-4

Cyclic B-Ketoesters: Synthesis and Reactions
M.A. Metwally and E. G. Sadek (Authors)
2010. 978-1-61668-282-8

Wet Electrochemical Detection of Organic Impurities
F. Manea, C. Radovan, S. Picken and J. Schoonman (Authors)
2010. 978-1-61668-661-1

Boron Hydrides, High Potential Hydrogen Storage Materials
Umit B. Demirci and Philippe Miele (Editors)
2010. 978-1-61668--362-7

Quantum Frontiers of Atoms and Molecules
Mihai V. Putz (Editor)
2010. 978-1-61668-158-6

Molecular Symmetry and Fuzzy Symmetry
Xuezhuang Zhao (Author)
2010. 978-1-61668-528-7

Molecular Symmetry and Fuzzy Symmetry
Xuezhuang Zhao (Author)
2010. 978-1-61668-375-7

Physical Organic Chemistry: New Developments
Karl T. Burley (Editor)
2010. 978-1-61668-469-3

Tetraazacyclotetradecane Species as Models of the Polyazacrown Macrocycles
Ryszard B. Nazarski (Author)
2010. 978-1-61668-487-7

Wet Electrochemical Detection of Organic Impurities
F. Manea, C. Radovan, S. Picken and J. Schoonman (Authors)
2010. 978-1-61668-491-4

Chemical Crystallography
Bryan L. Connelly (Editor)
2010. 978-1-61668-513-3

Chemical Reactions in Gas, Liquid and Solid Phases: Synthesis, Properties and Application
G.E. Zaikov and R.M. Kozlowski (Editors)
2010. 978-1-61668-671-0

Handbook of Inorganic Chemistry Research
Desiree A. Morrison (Editor)
2010. 978-1-61668-712-0

Molybdenum and Tungsten Cofactor Model Chemistry
Carola Schulzke and Prinson P. Samuel (Authors)
2010. 978-1-61668-750-2

Electrochemical Oxidation and Corrosion of Metals
Elena P. Grishina and Andrew V. Noskov (Authors)
2010. 978-1-61668-826-4

Molybdenum and Tungsten Cofactor Model Chemistry
Carola Schulzke and Prinson P. Samuel (Authors)
2010. 978-1-61668-828-8

Tetraazacyclotetradecane Species as Models of the Polyazacrown Macrocycles
Ryszard B. Nazarski (Authors)
2010. 978-1-61668-900-1

Chemical Reactions in Gas, Liquid and Solid Phases: Synthesis, Properties and Application
G.E. Zaikov and R.M. Kozlowski (Editors)
2010. 978-1-61668-906-3

Chemistry Research and Applications

Molybdenum and Tungsten Cofactor Model Chemistry

Carola Schulzke and Prinson P. Samuel

Nova Science Publishers, Inc.
New York

Copyright © 2010 by Nova Science Publishers, Inc.

All rights reserved. No part of this book may be reproduced, stored in a retrieval system or transmitted in any form or by any means: electronic, electrostatic, magnetic, tape, mechanical photocopying, recording or otherwise without the written permission of the Publisher.

For permission to use material from this book please contact us:
Telephone 631-231-7269; Fax 631-231-8175
Web Site: http://www.novapublishers.com

NOTICE TO THE READER

The Publisher has taken reasonable care in the preparation of this book, but makes no expressed or implied warranty of any kind and assumes no responsibility for any errors or omissions. No liability is assumed for incidental or consequential damages in connection with or arising out of information contained in this book. The Publisher shall not be liable for any special, consequential, or exemplary damages resulting, in whole or in part, from the readers' use of, or reliance upon, this material.

Independent verification should be sought for any data, advice or recommendations contained in this book. In addition, no responsibility is assumed by the publisher for any injury and/or damage to persons or property arising from any methods, products, instructions, ideas or otherwise contained in this publication.

This publication is designed to provide accurate and authoritative information with regard to the subject matter covered herein. It is sold with the clear understanding that the Publisher is not engaged in rendering legal or any other professional services. If legal or any other expert assistance is required, the services of a competent person should be sought. FROM A DECLARATION OF PARTICIPANTS JOINTLY ADOPTED BY A COMMITTEE OF THE AMERICAN BAR ASSOCIATION AND A COMMITTEE OF PUBLISHERS.

LIBRARY OF CONGRESS CATALOGING-IN-PUBLICATION DATA

Available upon Request
ISBN: 978-1-61668-750-2

Published by Nova Science Publishers, Inc. ✦ *New York*

Contents

Preface		xi
Chapter 1	Introduction	1
Chapter 2	Early Dithiolene Chemistry	9
Chapter 3	Modeling the Sulfite Oxidase Family	11
Chapter 4	Modeling the Xanthine Oxidase Family	19
Chapter 5	Modeling the DMSO Reductase Family	23
Chapter 6	Modeling the Three Families of Tungsten Enzymes	31
Chapter 7	OAT and PCET Reactions	37
Chapter 8	Future Perspectives	45
Abbreviations		49
References		51
Index		59

PREFACE

Molybdenum is crucially important for almost all organisms ranging from ancient single cell microorganisms to modern human being. This is somehow unusual since molybdenum is the only second row transition element known to have a role in vital enzymatic processes. Its higher homologue tungsten is the only third row transition element of biological importance and it is used in similar enzymes (with the exception of nitrogenase) for analogous reactions especially if molybdenum is not available in a specific habitat. The reactions catalyzed by molybdenum and tungsten cofactors involve two electron oxidation or reduction usually accompanied by an oxygen atom transfer from water to substrate or vice versa as part of the carbon, nitrogen and sulfur metabolism. The coordination environment of molybdenum and tungsten in these cofactors consists of one or two molecules of molybdopterin, which is a unique ligand in molybdenum and tungsten cofactors, establishing the coordination to the metal through a dithiolene function, and a varying combination of oxo, sulfido, hydroxo, water and/or amino acid residue ligands. Since the discovery of the molybdenum and tungsten cofactors it has been the aim of several bioinorganic groups to understand their structure and functionality by model chemistry. This was fueled in 1982 by the findings of Rajagopalan and co-workers that the unique molybdopterin ligand is part of these cofactors. From now on molybdenum and later tungsten dithiolene chemistry became an important part of this model chemistry. Bioinorganic chemists' attempts in this respect have afforded so far quite a number of molybdenum and tungsten complexes being structural or functional models or both. Some but not all of these compounds include dithiolene ligands. The coordination of a synthetic molybdopterin to molybdenum or tungsten, however, is still elusive. This as well as the still ongoing discovery of new molybdenum and tungsten active sites complimented with advances in the analytical and theoretical fields poses a non-ebbing challenge to bioinorganic chemists working in this field.

Chapter 1

INTRODUCTION

The biochemical exceptionality of molybdenum and tungsten lies in the fact that they are the only second and third row transition elements of substantial biological significance and that they are indeed utilized by almost all organisms ranging from archaea, the single celled ancient microorganisms, to modern human being. Chemically very similar, both metals are used in nearly identical enzymes with the exception of nitrogenase for which no tungsten analogue is known. The successive development of tungsten and molybdenum enzymes is most likely related to the changing bio-availability of both elements during the course of evolution.[1] It is believed that life arose in hot, sulfur rich anaerobic habitats in which the biochemical availability of molybdenum is extremely low due to *the infinitesimal solubility of molybdenum sulfides*. Under these conditions tungsten, however, is able to stay in oxidation state six and form *bis*-anionic and therefore soluble $[WO_xS_{x-n}]^{2-}$ (n = 0 – 4) species subsequently being bio-available to the first microorganisms. In addition it was argued that the higher solubility of low valent tungsten sulfides in comparison to molybdenum sulfides in water made them easier available in the hot anaerobic environment of the early earth.[2] When earth's crust was cooled and photosynthetic organisms started to generate oxygen, this led to cooler habitats with lower sulfur concentration and an oxidizing atmosphere and molybdenum became easier available to the biological system compared to tungsten due to its generally higher abundance. Interestingly molybdenum is the most abundant transition element in the modern ocean, even though its abundance in earth's crust is comparably low.[3] This high abundance is facilitated by the exceptionally good water solubility of molybdate $[MoO_4]^{2-}$ ions, which are being liberated during oxidative weathering of continental crust containing on average of 1 to 2 p.p.m. molybdenum.[4-5] The increasing availability of soluble molybdenum species resulted in its cumulative incorporation into the active sites of several enzymes during the course of evolution, while the ancient organisms (i.e. mainly archaea), that still live in habitats resembling the conditions of the early earth, continued to utilize tungsten. In conclusion the evolution from tungsten to molybdenum enzymes is based on the change from scarcity to availability of molybdenum, which is supported by new findings on the delay in evolution of eukariots from 2.7 to 0.7 billion years ago[3]. It has been proposed, based on experimental results, that the anaerobic environment prevailed in the early earth caused the scarcity of molybdenum in sea water and consequently slowed down the development of nitrogen fixing bacteria. Since eukaryotes are unable to fix nitrogen themselves, the unavailability of nitrogen fixing prokaryotes resulted in a delay of evolution of the former during this geological time period. Besides sheer abundance being the reason for today's overwhelming use of molybdenum in modern organisms it has been proposed that the change from molybdenum to tungsten during evolution was in addition based on functionality, mainly related to the redox potential behavior upon temperature change.[6] It seems as if molybdenum provides a functional advantage by being influenced to a lesser degree by temperature fluctuation, further driving evolution to switch from tungsten to molybdenum. This issue, however, is still being explored.

In addition to their bio-availability, naturally, the chemical versatility of molybdenum and tungsten were fundamental for their utilization in biological systems.[2] They are redox active between the oxidation states IV and VI, with the availability of intermediate oxidation state V facilitating the proton coupled electron transfer (PCET) being part of the fundamental reactions they catalyze. Molybdenum (with exception of nitrogenase) and tungsten enzymes catalyze reactions of the general type:

$$X + H_2O \rightleftharpoons XO + 2H^+ + 2e^- \qquad (1)$$

In which X and XO are the generalized acceptor and donor molecules respectively. In this type of reactions, water acts as either a source or sink of oxygen.[7] Since the substrate and product in this reaction differ only by an oxygen atom (O^{2-}), it has been termed oxo transfer reaction and the enzymes catalyzing this type of reactions are called oxo transferases. All the molybdenum and tungsten enzymes except nitrogenase belong to this broad class. However, the term oxo transferase does not bear a mechanistic implication in every case, even though there are examples like the reduction of DMSO to DMS and oxidation of sulfite to sulfate which can be viewed as actual oxygen atom transfer reactions. In addition these reactions involve proton coupled electron transfer [8] allowing a net redox change between substrates and products and a stepwise recovery of the enzyme's active state. The mechanism of oxygen atom transfer is an active area of discussion which will be addressed in chapter 7.

Investigations exploring the active site structures of oxo transferases commenced more than three decades ago. Even though sulfur co-ordination to the molybdenum centers was known to chemists from early on, it was believed that the sulfur of cysteinate was solely responsible for this interaction until the year 1982 when Rajagopalan and co-workers proposed the sulfur atoms being part of a specific ligand system they called molybdopterin (MPT), and by this rebutted the former assumptions. In their remarkable work they isolated and characterized the di(carboxamidomethyl) derivatives of molydbdopterin by fluorescence and mass spectrometry.[9] The first proposed structure, based on the then available evidence, including pterin dithiolene chelation to the molybdenum in the cofactor of chicken liver sulfite oxidase is shown in figure 1.a. There was an unidentified group at the sixth position of the pterin nucleus. The figure shows that this first proposed form of molybdopterin consists of a bi-cyclic pterin ring with a side chain carrying an ene-dithiol group and a phosphate group with the dithiolene part being coordinated to molybdenum. Later, in 1987, the same authors proposed a refined model for the molybdenum cofactor in which molybdopterin is a 6-alkylpterin with a 4-carbon side chain containing an enedithiol on C-1' and C-2', a secondary alcohol on C-3', and a phosphorylated primary alcohol on C-4' (figure 1.b).[10] In 1995 Chan et al. published the crystal structure of an oxo transferase, the tungsten containing aldehyde ferredoxin oxidoreductase (AOR) from *Pyrococcus furiosus*, a hyperthermophilic archaeon.[11] The active site structure of this enzyme (figure 1.c) revealed the molybdopterin being bound to the metal via dithiolene sulfurs, exactly as proposed by Rajagopalan et al. much earlier. The now commonly accepted chemical structure of molybdopterin based on various crystallographic studies is shown in figure 1.d.

More than fifty molybdenum and tungsten enzymes are known today. Of all the enzymes carrying molybdenum, nitrogenase can be considered as an exceptional case. This is the only enzyme in which Mo does not bind to MPT but to an iron sulfur cluster, histidine and homocitrate.[12] All other molybdenum and tungsten enzymes are associated with this unusual ligand system and they are mononuclear. "Molybdopterin" refers only to the organic part of the active sites and it does not include molybdenum as its name may suggest. Consequently molybdopterin can be associated with both molybdenum and tungsten enzymes. Because all this might be confusing different names have been proposed throughout the literature as pyranopterindithiolate[7], pterin-dithiolene[13], pterin-ene-dithiolate[14] or even tungstopterin when bound to tungsten[15]. However, throughout this book the original term molybdopterin will be used referring to the tri-cyclic organic ligand shown in figure 1.d. Molybdopterin is coordinated to the metal center via its dithiolene function. Depending on the type of enzyme, the number of coordinated molybdopterin ligands can be one or two. In eukaryotes, molybdopterin bears a terminal phosphate group at the pyrane ring as shown in figure 1.d. In the case of prokaryotes it carries in addition a nucleotide at the end of a phosphate chain, which can be cytosine, guanosine, adenosine or inosine. The later are termed dinucleotide versions of molybdopterin because molybdopterin itself is structurally related to the nucleotides via its pterin part. Here the focus is molybdenum's and tungsten's closer coordination sphere in the oxo transferases and therefore mainly "mononucleotide" molybdopterin coordination and the related model chemistry will be discussed.

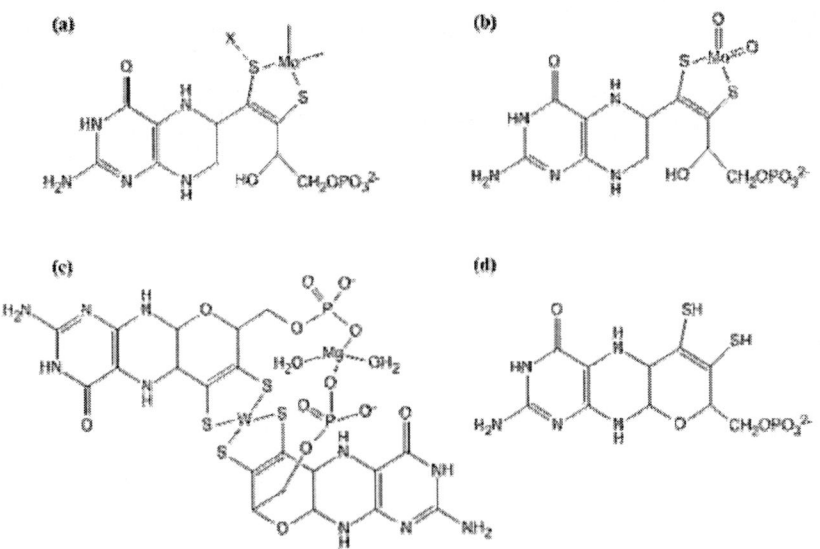

Figure 1. The development of insights into the structure of molybdopterin (MPT). a) the first proposed structure, b) the refined proposed structure, c) the first crystal structure of a MPT containing active site, d) today's commonly accepted structure of MPT.

Based on the geometrical and chemical structure of the oxidized active sites and sequence homologies, Hille has classified molybdenum and tungsten enzymes into different families.[2, 16] The molybdenum enzyme families are the xanthine oxidase family, the sulfite oxidase family and the DMSO reductase family. The two well known tungsten enzyme families are the aldehyde ferredoxin oxido reductase family and the formate dehydrogenase family. Each family is named after their most prominent member. The active site structures of each family of enzymes are depicted in figure 2.

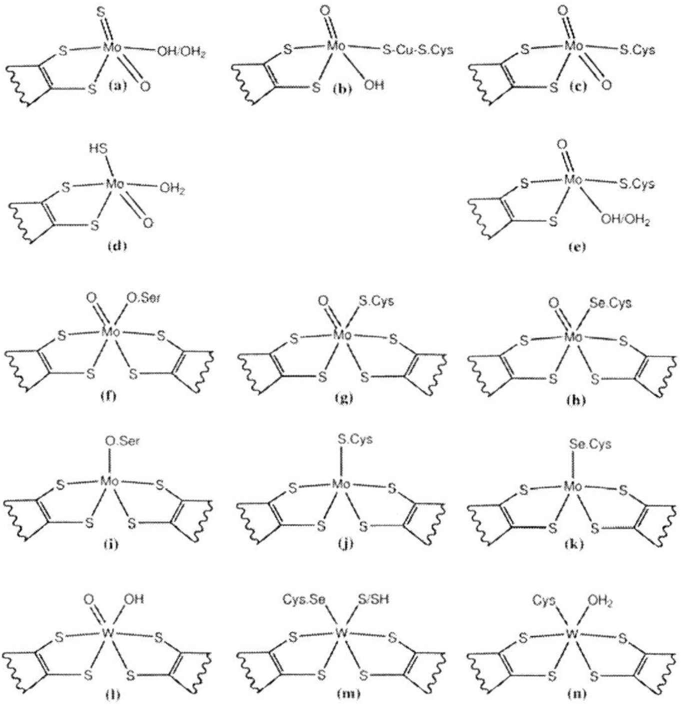

Figure 2. The active site structures of the different molybdenum and tungsten enzyme families as derived from crystal structures of individual enzymes. a) xanthine oxidase family (oxidized); b) unique active site of *Oligotropha carboxydovorans* CO dehydrogenase; c) sulfite oxidase family (oxidized); d) xanthine oxidase family (reduced); e) sulfite oxidase family (reduced); f)-h) DMSO reductase family enzymes with different coordinated amino acids (oxidized) (f: DMSO redcutase, g: dissimilatory nitrate reductase, h: formate dehydrogenase; i)-k) DMSO reductase family enzymes with different coordinated amino acids (reduced); l) aldehyde ferredoxin oxido reductase family: m) formate dehydrogenase family; n) acetylene hydratase.

The xanthine oxidase family of enzymes has an (MPT)Mo^{VI}OS(OH) core (figure 2.a) in the oxidized state whereas the structurally somehow similar sulfite oxidase family has an (MPT)Mo^{VI}O$_2$(S-Cys) core (figure 2.c). While both afore mentioned classes of molybdenum enzymes carry only one MPT equivalent in their active sites, the DMSO reductase family is an exception with its

(MPT)$_2$Mo^{VI}O(X) core possessing two MPT equivalents (figures 2.f-2.h). The group X can be a serine, an aspartate (not shown), a cysteine, a selenocysteine, a hydroxide (not shown) or a water molecule (not shown) with the amino acid residues being the most common ligands. Among the tungsten enzymes, the aldehyde ferredoxin oxidorecuctase family has the core structure (MPT)$_2$W^{VI}O(OH) (figure 2.l) analogous to the molybdenum center of arsenite oxidase. The second class of tungsten enzymes, the formate dehydrogenase family carries an oxidized core of the type (MPT)$_2$W^{VI}O(X) where X= S-Cys (not shown) or Se-Cys (figure 2.m). Both families show resemblance to the DMSO reductase family of molybdenum enzymes. The third family of tungsten dependent enzymes comprises only of a single member: the very unusual acetylene hydratase (figure 2.n), catalyzing the hydration of acetylene to acetaldehyde.[17]

The structural information about active sites of enzymes obtained by advanced analytical and theoretical means fueled the interest of bioinorganic chemists to explore the area of molybdenum and tungsten dithiolene chemistry in detail as these compounds would serve as model complexes for the molybdenum and tungsten cofactors in oxo transferases. The importance of model chemistry lies in the possibility of fine-tuning the ligand structure and coordination environment and by this directly paving the way for a trial and error approach towards understanding nature's choices with respect to coordination environment, ligand systems and active site metal. The versatility of oxo transferases in general but substrate specificity of the individual enzyme at the same time is still inspiring the bioinorganic community to develop even more accurate model systems. Another important aspect of model chemistry is the overcoming of analytical limitations associated with the complicated natural systems comprising the molybdenum and tungsten cofactors. The limitations in enzymatic active site characterization are based on the presence of heavy molybdenum and tungsten centers in varying oxidation states which might prevent a clear and unambiguous EXAFS evaluation and on the neighboring Fe-S clusters, heme irons or flavins having strong absorbing chromophores that may obscure electronic transitions of the active site metals.[18] The EXAFS studies are in addition limited in the determination of exact M-S bond length due to their multiple presence (2-4) in the same coordination sphere. The smallest resolvable difference by EXAFS for metal-ligand bonds involving similar ligands is >0.1 Å and by this equal to the uncertainty associated with data obtained from protein crystallography of most of the molybdenum and tungsten centers.[19,20] But most of all the instability of the isolated cofactors constitutes a major problem for thorough investigations. So the development of apt synthetic models mimicking the structural and chemical aspects of molybdenum and tungsten cofactors are essential for understanding the subtle and substantial variations of the natural active sites. Bioinorganic chemists' attempts in this respect have afforded so far numerous complex structures of molybdenum and tungsten, either with or without the ene-1,2-dithiolate (= dithiolene) function. The constantly increasing available information about the active sites of oxo transferases coined with their observed distinct catalytic functions continues to present inorganic chemists with several synthetic challenges to create exact mimics in order to unveil structure reactivity relationships. Complimented with advances in the analytical and theoretical fields and the discovery of new and distinct enzymes containing molybdenum or tungsten in their active sites, model studies allow the understanding of the function of metal, molybdopterin and the additional ligands to grow continuously, though to reach a detailed and complete comprehension of oxo transferases is still one of the big challenges in bioinorganic chemistry.

During the early years of dithiolene model chemistry, the $M^{IV}O/M^{VI}O_2$ couple was studied extensively as chemical analogues of molybdenum and tungsten cofactors due to its highly frequent occurrence in conventional inorganic molybdenum and tungsten chemistry. Recent structural information about the enzymes obtained by crystallographic methods prompted employment of the much more elusive $Mo^{IV}/Mo^{VI}O$ couple as model system with a much better resemblance of the actual active sites. The group of Richard H. Holm at Harvard was successful in developing the later mentioned class of complexes in an impressive variability. Nevertheless, even compounds of the former class hold justification for being valuable model systems with respect to certain aspects of the enzyme's active sites and are still being investigated today. In addition to this, present synthetic challenges involve the development of either dithiolene or *non*-dithiolene complexes with uncommon cores like $Mo^{VI}OS$, $Mo^{IV}O(SH)$, $Mo^{VI}O(OH)(S-Cu-S)$, $Mo^{IV}O(OH)_2(S-Cu-S)$, $Mo^{VI}SSe$ and $W^{VI}SSe$. Finally, a complete model of any enzyme with an exact mimic of the MPT ligand is remaining as a huge synthetic challenge. This book aims to focus on the development of the distinct model chemistry leading from its very beginning based on early and incomplete information about molybdenum's and tungsten's coordination environment to the elaborate recent approaches towards understanding the molybdenum and tungsten oxo transferases in detail. This includes a brief description of the infant stages of dithiolene chemistry of molybdenum and tungsten and its subsequent emergence as a branch of bioinorganic chemistry. Towards the end of the book oxo-transfer catalysis carried out using the model complexes will be discussed briefly finishing with an outlook into future developments in this fascinating field.

Chapter 2

EARLY DITHIOLENE CHEMISTRY

The research on metal related dithiolene chemistry was fuelled in the early 1960's with the pioneering works from the research groups of Schrauzer at Munich, Davison and Holm at Harvard and Gray at Columbia. A detailed and comprehensive analysis of the early stages of this field is available elsewhere in the literature.[21,22] Most of the early work was based on exploring the main aspects like geometry, unique electronic structure and one electron redox activity. While bisdithiolene complexes of late transition metals exhibit square planar geometry, some of the trisdithiolene complexes have an unusual trigonal prismatic geometry instead of the widely known octahedral geometry of six coordinate complexes, which minimizes ligand interactions. It was this unexampled geometry shown by certain dithiolene complexes that triggered the early investigations in this field. The earliest reports of structurally characterized homoleptic bis-dithiolene complexes cover $[Ni(mnt)_2]^{2-}$ and $[Ni(S_2C_2Ph_2)_2]$[23-25] and the oldest tris-dithiolene complexes exhibiting trigonal prismatic geometry are $[Re(S_2C_2Ph_2)_3]$, $[Mo(edt)_3]$ and $[V(S_2C_2Ph_2)_3]$[26-29] (figure 3). The distinct structures of homoleptic bis- and tris- dithiolene complexes have been summarized by Beswick et al. [30]. In addition the unique electronic structures of the intensely colored transition metal dithiolene complexes have been extensively studied by spectroscopic, magnetic and theoretical means and the *non-innocence* character of the dithiolene ligands has been widely documented.[31-35] The chemical and electrochemical, to some extent delocalized one electron redox reactivity of dithiolene complexes of transition metals has initiated numerous studies in the early stages of dithiolene chemistry.[22,36] The different possible bonding characteristics of dithiolenes using their neutral dithioketone or dianionic dithiolate or even mixed radical forms influence the formal oxidation state of the metal for a given metal-ligand stoichiometry. So the actual structure of a bis-dithiolene complex will be a resonance hybrid of these limiting structures and the related redox chemistry is neither purely metal based nor purely ligand based. At the present scenario dithiolene chemistry has found its place in different applied fields ranging from material science to bioinorganic chemistry. The importance of dithiolene complexes in bioinorganic chemistry was disclosed by the pioneering work of Rajagopalan and co-workers in the early 1980's when they described the nature of the molybdopterin bound to the active site metal in molybdenum and tungsten cofactors. The much earlier dithiolene research based on curiosity with respect to the unique properties of dithiolene complexes and its further development towards material science was responsible for the availability of quite good model ligands for MPT as soon as its dithiolene character was revealed. The ground work in dithiolene chemistry allowed the immediate and convenient use of this type of ligands in bioinorganic model chemistry. Since then, numerous dithiolene complexes relevant for the molybdenum and tungsten cofactors in oxo transferases have been synthesized and extensively studied for their structural, spectroscopic, magnetic and redox characteristics.

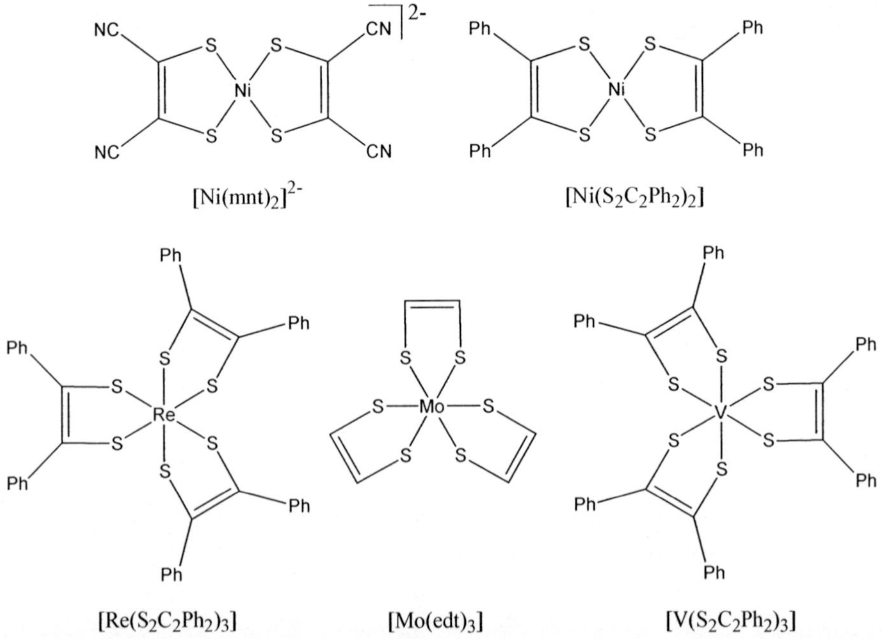

[Ni(mnt)$_2$]$^{2-}$ [Ni(S$_2$C$_2$Ph$_2$)$_2$]

[Re(S$_2$C$_2$Ph$_2$)$_3$]   [Mo(edt)$_3$]   [V(S$_2$C$_2$Ph$_2$)$_3$]

Figure 3. The earliest reports of structurally characterized homoleptic bis-dithiolene and tris-dithiolene complexes.

Chapter 3

MODELING THE SULFITE OXIDASE FAMILY

The sulfite oxidase (SO) family of enzymes comprises the sulfite oxidases and assimilatory nitrate reductases. Sulfite oxidases catalyze the transformation of sulfite to sulfate accompanied by a change of molybdenum's oxidation state from VI to IV in the reductive half reaction of the catalytic cycle (figure 4). The catalytic cycle is completed by re-oxidation of molybdenum, first to Mo^V and then to Mo^{VI}, by the electron transfer to the physiological oxidant cytochrome *c* via a *b*-type cytochrome site present in the enzyme.[37-40] Nitrate reductases are actually catalyzing the electronically reverse type of reaction executing the reduction of nitrate to nitrite with a formal oxidation change at molybdenum from IV to VI in the oxidative half reaction. The earliest attempts to investigate the coordination sphere of molybdenum in sulfite oxidases and nitrate reductases were documented by Cramer et al..[41,42] It was reported that $Mo^{VI}O_2$ and $Mo^{IV}O$ units were present in the oxidized and reduced forms of these enzymes, respectively. For oxidized sulfite oxidase the EXAFS analysis revealed two oxygen atoms at a distance of 1.68 Å to molybdenum and two or three sulfur atoms at 2.41 Å, changing to only one oxygen at 1.69 Å, and three sulfur atoms at 2.38 Å upon reduction. In case of the oxidized state of the assimilatory nitrate reductase, EXAFS results showed that molybdenum carried two terminal oxygen atoms at 1.71 ±0.03 Å as well as two or three sulfur atoms at 2.44 ± 0.03 Å. One single terminal oxygen at 1.67 ± 0.03 Å and a set of sulfurs at 2.37 ± 0.03 Å were found upon full reduction by NADH. Similar information was obtained by George et al. by X-ray absorption spectroscopy studies of the Mo^{IV}, Mo^V and Mo^{VI} oxidation states of SO.[43] These studies indicated that the Mo^{VI} oxidation state possesses two terminal oxo (M=O) and approximately three thiolate-like (Mo-S-R) ligands, unaffected by changes in pH and chloride concentration. The Mo^{IV}

and Mo^V oxidation states were found to carry one oxo ligand, one Mo-OX (most probably Mo-OH) and two to three sulfur ligands. X-ray absorption spectroscopy at the molybdenum and sulfur K-edges was carried out for the oxidized Mo^{VI} active sites of wild-type and cysteine 207 to serine mutant human sulfite oxidases.[44] The wild-type enzyme was found to possess two terminal oxygen ligands to molybdenum at 1.71 Å and three Mo-S distances of 2.41 Å whereas in the mutant one sulfur ligand was replaced by an oxygen ligand. With this it was proven that the amino acid residue of cysteine 207 was a ligand to molybdenum in the wild type. The crystal structure of chicken liver sulfite oxidase was published by Kisker et al. in 1997.[45] The structure showed that the active site molybdenum is five-fold coordinated by one oxo group, three sulfur ligands (two from the molybdopterin and one from the cysteinate ligand) and one water or hydroxo ligand. A little ambiguity was left in the report since the protein was purified in its fully oxidized form [Mo^{VI}/Fe^{III}] but molybdenum was found to be a mono oxo species in contrast with the afore mentioned EXAFS results and resonance Raman studies.[46] Nevertheless, subsequent experiments proved beyond doubt that the SO family of enzymes is characterized by a $Mo^{VI}O_2$ group in the oxidized form with a molybdopterin ligand coordinated through its dithiolene function and the chicken liver SO has been characterized to bear one cysteinate ligand at the molybdenum center.[47-49] The synthetic requirement for accurately modeling an enzyme of the sulfite oxidase family is to combine a $Mo^{VI}O_2$ core with three thiolate donors, two of which being part of a dithiolene ligand and the remaining an analogue of cysteine. In addition, one should consider that the three thiolate donors in SO are lying in an equatorial plane. The five atoms around the metal constitute a square pyramidal geometry with one oxo ligand at the apical position.

A large number of synthetic analogues has been reported for the SO family of enzymes. Many of them are structurally characterized and screened for their catalytic oxo-transfer properties. Most of the earlier developed model complexes for the SO family of enzymes were not able to correctly resemble the coordination environment with three thiolate and two terminal oxo groups in their oxidized form. However, the bioinorganic model chemistry in this area was developed with an explicit desire to obtain reactivity analogues. Similar to the catalytic reaction in the enzymes, the oxo-transfer catalysis by analogous complexes included the $Mo^{VI}O_2/Mo^{IV}O$ redox couple. Unfortunately the formation of a $Mo^V{}_2O_3$ core (figure 5) by reaction of both species with each other (comproportionation) inhibits any further activity and obstructs the catalytic cycle. In an attempt to prevent this deactivation by dimerization, Holm et al. synthesized $MoO_2(L-NS_2)$, (figure 6.a) where (L-NS$_2$) is a sterically bulky ligand preventing direct contact between the two metal centers at both ends of a single catalytic turn over.[50,51]

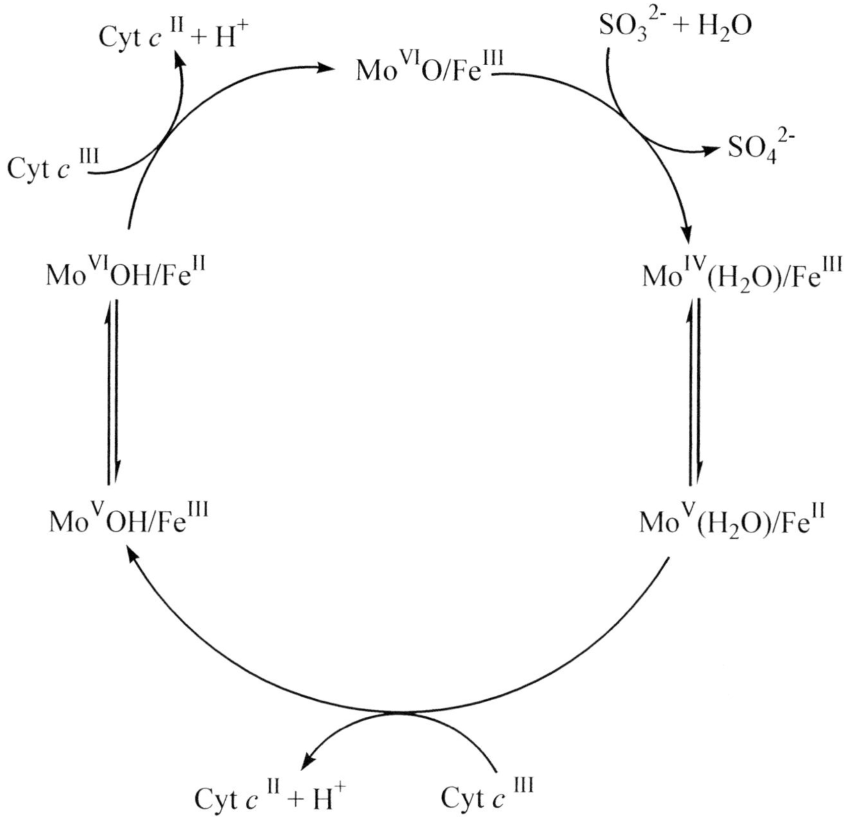

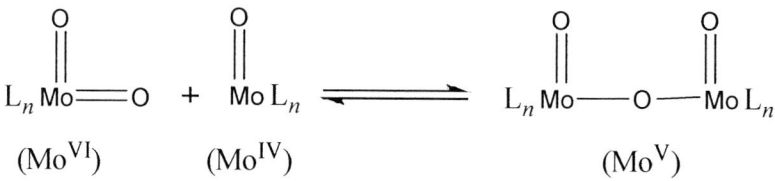

Figure 4. The proposed catalytic cycle of sulfite oxidase. Fe is the mediator cytochrome b for the electron transfer between the molybdenum active site and cytochrome c.

$$L_n Mo{=}O \;+\; O{=}Mo L_n \;\rightleftharpoons\; L_n Mo{-}O{-}Mo L_n$$
$$(Mo^{VI}) \qquad\qquad (Mo^{IV}) \qquad\qquad\qquad (Mo^{V})$$

Figure 5. Formation of the Mo$^V_2$O$_3$ core.

This complex was indeed efficiently catalyzing the oxygen atom transfer from DMSO onto PPh$_3$ to form DMS and PPh$_3$O, but catalytic turnover caused an unforeseen reorientation of the ligand allowing the formation of the Mo$_2$O$_3$ core anyway. After the paramount relevance of these steric factors in ligand systems had been demonstrated by Holm, as a first example of a coordinatively saturated molybdenum complex reacting readily and reversibly with organophosphines at room temperature in water or methanol, the Cervilla group synthesized [NH$_4$]$_2$[MoO$_2$(O$_2$CC(S)Ph$_2$)$_2$]·2H$_2$O (figure 6.b).[52] Both terminal oxo ligands at the molybdenum center of this complex are sterically shielded by one phenyl ring from the ligand in the direction of a potential Mo-O-Mo bond preventing dimerization. A similar complex with respect to steric and catalytic properties, incorporating two terminal oxo groups and two sulfur atoms from a sterically crowded ligand system was the six-fold coordinated Mo^{VI}O$_2$(Bu^{t}L-NS)$_2$ (figure 6.c). The reduced form of this complex, Mo^{IV}O(But-NS)$_2$, is a stable five-fold coordinated system.[53,54]

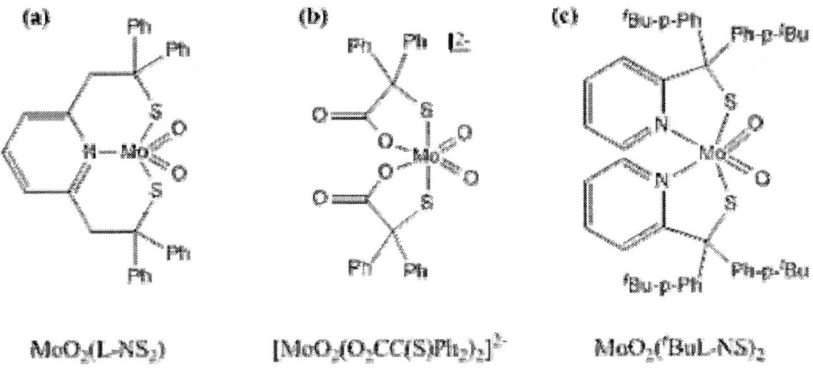

Figure 6. Functional model complexes for the sulfite oxidase family of molybdenum enzymes with sterically bulky ligands.

The regeneration of the sulfite oxidase's active center from Mo^{IV}O to Mo^{VI}O$_2$ is achieved through a Mo^V transient state as emphasized by EPR studies.[55] The oxygen atom transfer facilitated by the interconversion of Mo^{VI}O$_2$ and Mo^{IV}O of model compounds has been documented widely throughout the literature. But the first model complex which accomplished the regeneration of Mo^{VI}O$_2$ from Mo^{IV}O through two distinguished one electron transfer steps, accurately resembling the natural system, was Tp*Mo^{VI}O$_2$(SPh); [Tp* = hydrotris(3,5 dimethyl- 1-pyrazolyl)borate anion].[56] This complex is another example of the class of compounds which is noted for the presence of sterically highly demanding ligand systems able to minimize or prevent comproportionation of reduced and fully oxidized species. In DMF or MeCN oxygen atom abstraction by PPh_3 from Tp*MoO$_2$(SPh) leads to the formation of Tp*Mo^{IV}O(SPh) or weakly solvated Tp*Mo^{IV}O(SPh)(solvent). It can be trapped as mononuclear species in pyridine Tp*Mo^{IV}O(SPh)(py) or in CH_2Cl_2 Tp*Mo^{V}OCl(SPh). In contrast, it forms dinuclear [Tp*Mo^{V}O(SPh)]$_2$O in dry toluene. The reaction takes yet another course in wet THF or toluene: the presence of small amounts of water results in the formation of Tp*Mo^{V}O(OH)(SPh) according to the reaction:

$$Tp*Mo^{IV}O(OH_2)(SPh) + Tp*Mo^{VI}O_2(SPh) \rightarrow 2\ Tp*Mo^VO(OH)(SPh) \quad (2)$$

Tp*Mo^{V}O(OH)(SPh) can be oxidized quantitatively to Tp*Mo^{VI}O$_2$(SPh). The different reactions observed are summarized in figure 7. Consequently, the Tp*Mo^{VI}O$_2$(SPh) complex is a catalyst for the oxidation of PPh_3 to PPh_3O. Oxygen isotope (^{18}O) tracing using labeled H_2O shows that the oxygen in PPh_3O stems from water. This was the first model system which displayed the full catalytic cycle proposed for oxidizing molybdenum enzymes featuring the $[Mo^{VI}O_2]^{2+}$ resting state in resemblance to SO.[57]

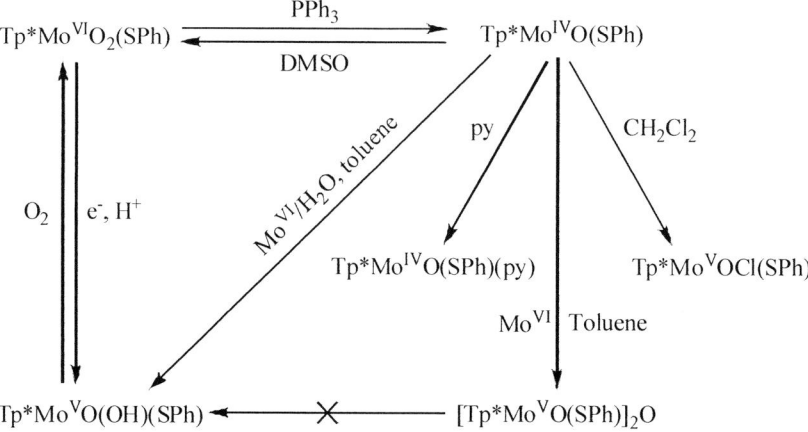

Figure 7. Catalytic oxo transfer activity, regeneration of Mo(VI)O$_2$ from Mo(IV)O through two distinguished one electron transfer steps and distinct behavior in different solvents.

With a more structural than functional focus on the SO enzymes active sites cis-, trans-[Mo^{V}O(L-N$_2$S$_2$)(SR)] (L-$N_2S_2H_2$ = N,N'-dimethyl-N,N'bis(mercaptophenyl)ethylenediamine; R = CH_2Ph, CH_2CH_3, and p-C_6H_4−Y (Y = CF_3, Cl, Br, F, H, CH_3, CH_2CH_3, and OCH_3) have been developed. These compounds contain three thiolate donors bound to molybdenum in the equatorial plane (figure 8). Notably, these are the first structurally characterized mononuclear molybdenum compounds with three thiolate donors.[58] Even though not exactly reflecting the ligand environment of the active site molybdenum in SO, its coordination geometry places two adjacent S pπ orbitals parallel to the Mo=O bond, analogous to the orientation of the MPT ligand in sulfite oxidase, with the third S pπ orbital lying in the equatorial plane.

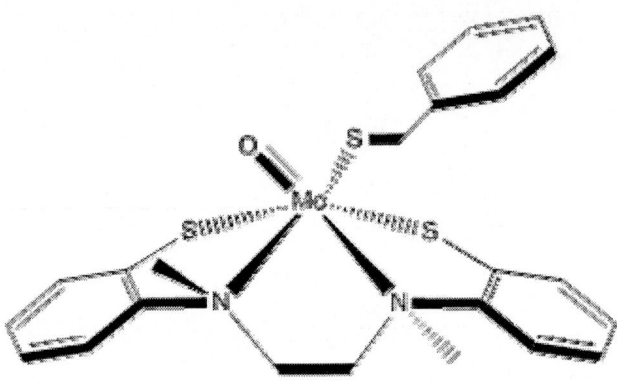

Figure 8. [Mo^{V}O(L-N$_2$S$_2$)(SCH$_2$Ph)] - three thiolate donors bound to molybdenum in the equatorial plane.

Closer coordination environment mimics of the SO family of enzymes need to bear exactly one dithiolene ligand at the molybdenum center. Such mono-dithiolene complexes were regarded as unusual and difficult to synthesize until recently but their relevance was underlined after the cofactor of SO was characterized crystallographically. Two very different synthetic pathways for realizing this type of complexes were described by the Holm group.[59] The first method leads to the formation of mono-dithiolene Mo^V complexes analogous to the high pH form of sulfite oxidase by starting with the *bis*-dithiolene complexes. For example, [Mo^{V}O(S$_2$C$_2$Me$_2$)$_2$]$^{1-}$ or [Mo^{V}O(bdt)$_2$]$^{1-}$ (bdtH$_2$ = 1,2 benzene-dithiol) is reacted with PhSeCl in a way that one of the ene-dithiolate ligands is replaced by two chlorides forming [Mo^{V}OCl$_2$(S$_2$C$_2$Me$_2$)]$^{1-}$ or [Mo^{V}OCl$_2$(bdt)]$^{1-}$ respectively. Upon treatment with a sterically hindered thiol, these complexes undergo ligand substitution reactions forming other monodithiolene Mo(V) complexes [MoO(2-AdS)$_2$(S$_2$C$_2$Me$_2$)]$^{1-}$, [MoO(SR)$_2$(bdt)]$^{1-}$ with R = 2-Ad or 2,4,6-Pr$^i_3$C$_6$H$_2$, and [MoOCl(SC$_6$H$_2$-2,4,6-Pr$^i_3$)(bdt)]$^{1-}$ (Ad = adamantyl). The latter complex presents the closest similarity to the active site structure of sulfite oxidase in its high pH form with one oxo, one thiol, one dithiolene and the chloro ligand, which is in place of the hydroxide (figure 9). The second synthetic and slightly more economic route leads to analogues of the fully oxidized form of SO. Treatment of [Mo^{VI}O$_2$(OSiPh$_3$)$_2$] with Li$_2$(bdt) in THF affords [Mo^{VI}O$_2$(OSiPh$_3$)(bdt)]$^{1-}$ which reacts with the bulky thiol 2,4,6-Pr$^i_3$C$_6$H$_2$SH in acetonitrile giving [Mo^{VI}O$_2$(SC$_6$H$_2$-2,4,6-Pr$^i_3$)(bdt)]$^{1-}$. The result is an excellent model for the oxidized form of SO with both of the oxo ligands, a thiolate in place of cysteinate and benzenedithiolene mimicking MPT.

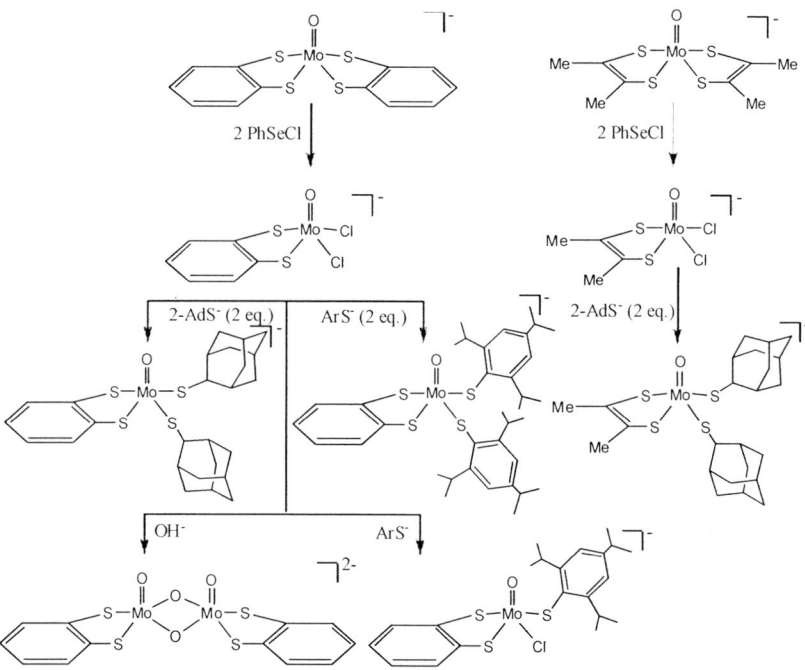

Figure 9. Synthesis of monodithiolene complexes from bis(dithiolene) complexes.

Chapter 4

MODELING THE XANTHINE OXIDASE FAMILY

The xanthine oxidase family comprises several enzymes which bear the $LMo^{VI}O(S)$ core at their active sites, where L is the molybdopterin ligand system. Depending on whether the enzyme is eukaryotic or prokaryotic and which reaction is catalyzed, L can be the mononucleotide form of the pterin cofactor (molybdopterin), molybdopterin cytosine dinucleotide, molybdopterin guanine dinucleotide or molybdopterin adenine dinucleotide.[16] Although known as xanthine oxidase (XO) for a long time, detailed and up-to-date knowledge of structure and function strongly suggests a referral to these enzymes by the more recent and more accurate name xanthine oxidoreducase (XOR). The reactions catalyzed by XO family enzymes usually involve cleaving an R-H bond and forming an R-O bond:

$$RH + H_2O \rightarrow ROH + 2 H^+ + 2e^- \qquad (3)$$

Consequently, they are called hydroxylases with respect to function and not oxo transferases. Many attempts have been made to explore the coordination environments of molybdenum in the XO family of enzymes, especially making use of x-ray absorption spectroscopy techniques.[41,60-62] The first molybdenum hydroxylase to be crystallographically characterized was the aldehyde oxidoreductase from *Desulfovibrio gigas*.[63-64] These studies suggested an overall square-pyramidal geometry around molybdenum and the composition of the active site to be $LMoO_2(OH_{(2)})$. Of the two terminal oxo groups, one is in the apical position and the other in an equatorial position with a close resemblance to the oxidized SO family structures. Unfortunately, the enzyme used in this initial crystallographic study was the inactive desulfo form lacking a terminal sulfido ligand. Electron density studies on the activated enzyme obtained by treating crystals with a resulfurating agent suggested that it was the apical oxo group that was replaced by sulfur to form Mo=S. But a single Mo=O group is supposed to be in the apical position and opposite to the vacant coordination site of a square-pyramidal coordination geometry, owing to its strong *trans* influence. The reverse assignment was quite unlikely from the chemical point of view. Eventually a freeze−quench magnetic circular dichroism spectroscopic study of the "very rapid" intermediate of xanthine oxidase supported indeed an apical position for the oxo ligand as in common model complexes.[65] Subsequent crystal structures of other enzymes in this family (CO dehydrogenase from the aerobic *Oligotropha carboxydovorans* and quinoline-2-oxidoreductase from *Pseudomonas putida*) also underlined the fact that it is the equatorial position that becomes sulfurated upon activation in molybdenum hydroxylases.[66-69] A notable and very unusual feature in the co-ordination sphere of molybdenum in CO dehydrogenase is the fact, that a copper ion occupies a position in the equatorial plane being connected to molybdenum by a sulfide bridge which is in place of the catalytically essential sulfur of other members of this enzyme family (figure 2.b).

The model chemistry of this family of enzymes is still in its infant stage, as the $Mo^{VI}OS$ core with additional dithiolene chelation is an uncommon motive in synthetic inorganic chemistry. In the enzymes the $Mo^{VI}OS$ center is believed to be converted to a $Mo^{IV}O(SH)$ species, following hydride transfer from the substrate to the sulfido group and this presents additional synthetic challenges.[18] After all, the fundamental importance of understanding the chemical role of bridging or terminal sulfur atoms in the catalytic turnover still challenges bioinorganic chemists working in this field. Early examples of simple synthetic analogues of the xanthine oxidase family of enzymes include oxothiomolybdates, $[MoO_{4-n}S_n]^{2-}$ [70], hydroxylamido complexes, $MoOS(ONR_2)_2$[71-72] and organometallic derivatives as $Cp*MoOS(CH_2SiMe_3)$[73]. One of the first truly comparative studies of analogue complexes and different species of milk xanthine oxidase was carried out by Wedd and co-workers.[74] Model complexes used in this work include $[MoOXL]^{1-}$ and $[MoO(XH)L]$, where X = O, S and LH_2 = N,N'-dimethyl-N,N'-bis(2-mercaptophenyl)- 1,2-diaminoethane (figure 10). Rapid Type 1, Rapid Type 2, and Slow centers of milk xanthine oxidase (named after characteristic EPR signals upon enzyme reduction by different methods) were compared to the synthetic species $[MoOXL]^{1-}$ and $[MoO(XH)L$ (X=O, S) by definition of the ^{95}Mo hyperfine matrices from multifrequency EPR spectra. Similarly, ^{33}S hyperfine matrices were extracted for the sulfide ligand of $[MoOSL]^-$ and for the mercapto ligand SH^- of $[MoO(SH)L]$ and compared with the ^{33}S coupling observed in a number of EPR signals of xanthine oxidase species. Detailed evaluation revealed [Mo^{V}O(SH)], [Mo^{V}O(SH)(OH)], and [Mo^{V}O(OH)] cores as being responsible for the Rapid Type 1, Rapid Type 2, and Slow EPR signals. Strong evidence for [Mo^{V}OS] as being the source of the very rapid signal was obtained in addition.

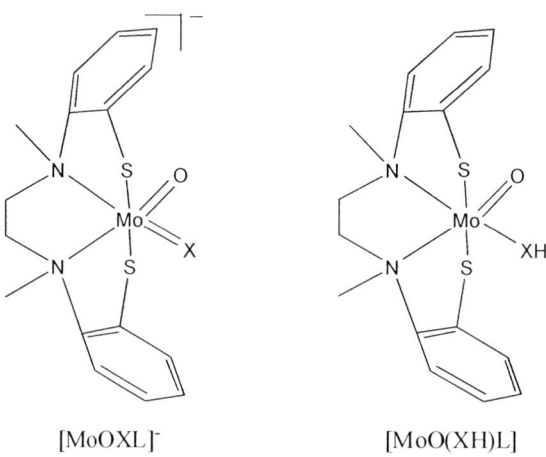

[MoOXL]$^-$ [MoO(XH)L]

Figure 10. Chemical structures of [MoOXL]$^-$ and [MoO(XH)L]; (X= S or O).

Recently synthetic model chemistry was advanced by Young and co-workers, as they were successfully utilizing oxygen atom transfer and sulfur atom transfer reactions in the preparation of mononuclear oxo-sulfido-molydenum(VI) complexes such as Tp*MoO(S){SP(S)Pr$^i_2$} [Tp* = hydrotris-(3,5-dimethylpyrazol-1-yl)borate], Tp'MoO(S){SP(S)R$_2$} [Tp' = hydrobis(3-isopropylpyrazolyl-1-yl)(5-isopropylpyrazolyl-1-yl)borate; R = Pr^i, Ph] and Tp^{iPr}MoOS(OAr) [Tp^{iPr} = hydrotris(3-isopropylpyrazol-1-yl)borate; OAr = phenolate] (figure 11).[75,76] As a notable feature both of the former cis-oxosulfidomolybdenum(VI) complexes are stabilized by a weak, intra-molecular Mo(O)=S⋯S=P interaction generating five membered rings. The bond order of the weak interaction was calculated to be approximately ⅓ but it still lengthens the Mo=S bond slightly. Interestingly and importantly this weak S⋯S interaction prevents reduction of molybdenum/oxidation of sulfur and the subsequent dimerization leading to Mo^{V}O(S-S)Mo^{V}O species, which is an unwanted but common feature and one of the major obstacles in the chemistry of Mo^{VI}OS. In contrast and as expected in the absence of the stabilizing S⋯S interaction the otherwise very similar Tp^{iPr}MoOS(OAr) dimerizes in solution.

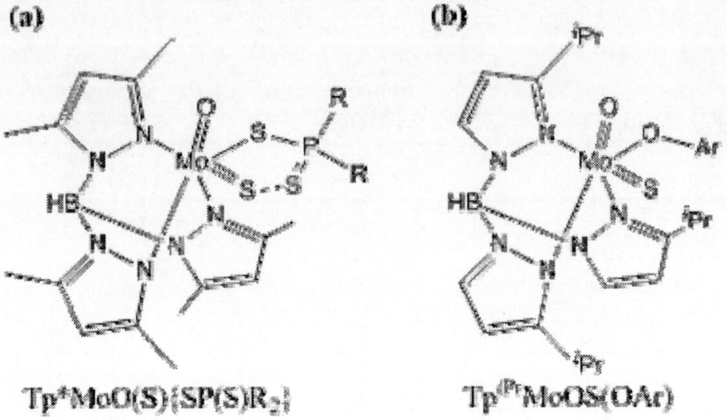

Figure 11. (a) Tp*MoO(S){SP(S)R$_2$} and (b) Tp^{iPr}MoOS(OAr).

Chapter 5

MODELING THE DMSO REDUCTASE FAMILY

The DMSO reductase (DMSOR) enyzme family is different from other molybdenum enzymes regarding the number of MPT ligands. Here two molecules of MPT in its dinucleotide form are bound to molybdenum. These enzymes have the general active site composition $L_2Mo^{VI}(O)R$ in the oxidized state and $L_2Mo^{IV}(R)$ in the reduced state (L = MPT and R = ligand most often contributed by the polypeptide). Compared to the other two molybdenum families of enzymes, the DMSOR family is large and diverse with respect to substrates but restricted to the prokaryotic regime. Crystal structures of several enzymes of the DMSOR family are known today. One of the first was the crystal structure of DMSO reductase from *Rhodobacter sphaeroides* revealing a mono-oxo molybdenum(VI) cofactor containing two molybdopterin guanine dinucleotides (MGD) and one serinate (O-Ser).[77] One of the MGDs exhibits different coordination modes to the molybdenum in the oxidized and in the reduced state. The change in MGD coordination from the Mo^{VI} species to the Mo^{IV} species is supposed to be crucial in the mechanism for substrate binding and reduction by this enzyme. In the crystal structure of oxidized trimethylamine *N*-oxide reductase (TMAOR) from *Shewanella massilia* molybdenum was found to be ligated by four sulfur atoms from the two MGDs, two oxo groups and the oxygen of Ser149 (serine) which would constitute an unfavorable seven-fold coordination.[78] The proposed structure faces several serious limitations such as anomalous bond lengths (e.g. the Mo−O(Ser149) bond length at 1.7 Å is far too short for an alcoholate coordination to molybdenum and the four Mo−S bonds are all suspiciously long, at 2.5, 2.6, 2.7, and 2.8 Å). All this indicates that the postulated molybdenum coordination structure by Czjzek et al. is chemically impossible. A re-examination of the proposed crystal structure by Mo-K edge EXAFS experiments of the molybdenum site of recombinant *Escherichia coli* trimethylamine *N*-Oxide reductase revealed substantial discrepancies between both studies.[79] The EXAFS data of the redox-cycled enzyme (reoxidation of reduced enzyme as part of the purification processes) indicates a single Mo=O ligand at 1.71 Å, four Mo−S ligands (from two MGDs) at 2.43 Å, and one Mo−O-Ser ligand at 1.83 Å just like the oxidized form of DMSO reductase. An explanation for the shortcomings in the crystal structure determination most likely is the presence of the enzyme in different oxidation states leading to a superposition of different forms.

The first reported crystal structure of dissimilatory nitrate reductase from *Desulfovibrio desulfuricans* also displays a similar ligand environment of the molybdenum center with cysteinate (S-Cys) as the amino acid residue connecting molybdenum and polypeptide. In the oxidized form of the enzyme, Mo^{VI} is coordinated by six ligands in the typical distorted trigonal prismatic geometry. Four ligands are provided by the two ene-dithiolate sulphur atoms of each MGD and the fifth ligand is Cys140 (cysteine). During catalytic turnover the sixth coordination position is occupied by an oxo ligand. The oxidized form will be transferred back to the square pyramidal desoxo Mo^{IV} form through a state of hydroxo/water co-ordination by subsequent proton and electron abstraction processes.[80] The X-Ray structural investigation of respiratory nitrate reductase (Nar) from *E. Coli* also shows a similar geometry only with aspartate (Asp) as the polypeptide ligand.[81] In contrast to the co-ordination environment seen in all other enzymes of the DMSOR family, that were structurally studied, another member, arsenite oxidase, of the same family bears a different coordination sphere at the molybdenum center. The crystal structure of arsenite oxidase from *Alcaligenes faecalis* shows a five-fold coordinated Mo^{IV}=O center with the other four co-ordination sites occupied by the usual four ene-dithiolate sulfur atoms of two MGDs.[82] The oxidized state is a six coordinate dioxo species. The most significant difference between arsenite oxidase and the other members of the DMSO reductase family is the absence of any covalent linkage between the protein and the molybdenum atom and the presence of an oxo ligand in the reduced state. In all other reported enzymes of this family, the molybdenum is coordinated by the side chain of a serine, aspartate, cysteine, or selenocysteine amino acid. In arsenite oxidase, the corresponding residue is an alanine, which is unable to develop a coordinative bond and consequently there is no direct connection between molybdenum and polypeptide. X-ray absorption spectroscopic studies on AO from *Alcaligenes faecalis* indicate that the Mo−S bonds shorten from 2.47 to 2.37 Å upon reduction with the physiological substrate. The studies also confirm that an oxo ligand at 1.70 Å is present in both oxidized and reduced forms of the enzyme and that the oxidized form has an additional Mo–O bond at 1.83 Å which is lost upon reduction.[83] This longer Mo-O distance is not conclusive with a second oxo ligand as found in the X-ray structure but the authors stress that the difference to the more typical Mo=O length of 1.70 Å is close to the resolution limit and suggest that this is either an oxo or a hydroxo ligand. Subsequently at present it is being discussed to consider arsenite oxidase as representing a fourth family of Mo enzymes.[84] Another enzyme, which is assumed to have a bis(MPT)Mo center, is selenate reductase. The active site of this enzyme from *Thauera selenatis* has been characterized by Mo, Se, and Fe K-edge X-ray absorption spectroscopy. It was found that the molybdenum site of the oxidized enzyme carries 3 to 4 sulfur ligands from two molybdopterins at 2.33 Å, one terminal oxo group at 1.68 Å and one Mo–O bond with an intermediate bond length of 1.81 Å. The reduced enzyme has a des-oxo Mo center, with nearly four sulfur ligands at 2.32 Å and possibly one Mo-O bond at 2.22 Å.[84] The enzyme was found to contain selenium in a reduced form. However, *T. selenatis* selenate reductase does not contain the SECIS (selenocysteine insertion sequence) and therefore the selenium is unlikely to be part of a selenocysteine. Selenate reductase most probably contains a six coordinate $Mo^{VI}O(OH)$ center with two MPTs in the oxidized form and a five coordinate $Mo^{IV}(OH)$ with two MPTs in the reduced form. It is therefore analogous to the active site structures that are characteristic for the DMSO reductase family of molybdenum enzymes without the molybdenum peptide bond. The selenium site of selenate reductase, however, still remains a puzzle until more detailed future spectroscopic and crystallographic investigations will be conducted with an unanimous outcome.

The above description demonstrates that model chemistry for the DMSO reductase family of enzymes can take two courses: firstly, modeling the $Mo^{IV}S_4R/Mo^{VI}(O)S_4R$ (R = Amino acid analogue) couple as in the case of structurally characterized DMSOR, TMAOR and NIR enzymes and secondly modeling the $Mo^{IV}(O)S_4/Mo^{VI}(O)_2S_4$ couple as in the case of AO. Most of the early developed model complexes come under the second class, even though at the time of their development the structure of AO was unknown. Bearing in mind the active site structures of the large majority of the DMSOR family enzymes it is clear that rather mono-oxo six co-ordinate Mo^{VI} compounds are required when modeling enzymes of this family but such complexes are rarely reported. The most frequently reported form in the early synthetic model chemistry was the *cis*-dioxo moiety MoO_2^{2+}.[85] At the same time, of the early mono-oxo Mo^{VI} compounds a large number was even seven-fold coordinated.[86-88] In 1988, Mondal et al. reported MoO(cat)(Sap) complexes (cat = catecholate (Cat^{2-}), naphthalene-2,3-diolate (Naphcat^{2-}), or 3,5-di-*tert*-butylcatecholate (DTBcat^{2-}) and Sap = the Schiff base dianion N-salicylidene-2-aminophenolate) as first examples of mononuclear MoO^{4+} centers with a six-fold coordination (figure 12).[89] Following this, several mono-oxo Mo^{VI} coordination complexes were reported and some of them were screened for understanding redox potentials and oxo-transfer properties.[90-95]

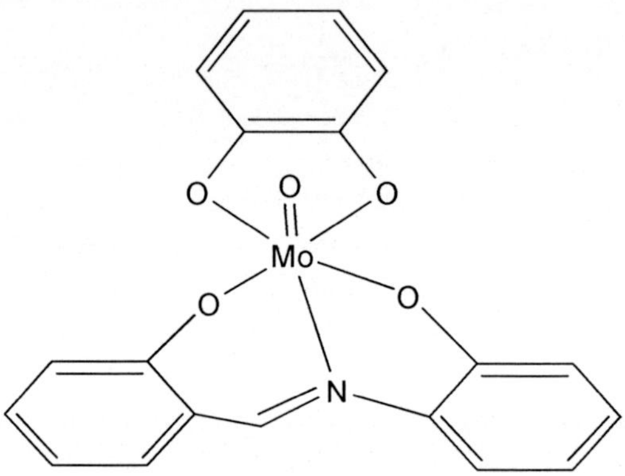

Figure 12. Chemical structure of MoO(Cat)(Sap).

The accessible molybdenum oxidation state V is assumed to act as mediator between the obligatory two-electron and one-electron redox systems and a detailed understanding of its properties is equally important as for the more commonly observed enzymatic oxidation states Mo^{IV} and Mo^{VI}. Nearly two decades prior to the publication of crystal structures of DMSO family enzymes Wedd and co-workers synthesized [Mo^{V}O(SPh)$_4$]$^-$ which can be considered as one of the very early developed model complexes in this series.[96] After that, several complexes having [Mo^{V}OS$_4$]$^-$ coordination have been synthesized and extensively studied theoretically and experimentally with respect to geometry, electronic structure, redox properties and bonding.[97-100] In an effort to identify the geometric control of the redox potential by serine ligation to the molybdenum center of DMSO reductase, Kirk, Basu and co-workers synthesized and isolated *cis*-[(L1O)Mo^{V}OCl$_2$] and *trans*-[(L1O)Mo^{V}OCl$_2$] complexes [L1OH = (3-*tert*-butyl-2-hydroxy-5-methylphenyl)bis(3,5-dimethylpyrazolyl)methane] whith the *cis* and *trans* notation referring to the position of the phenolato oxygen relative to that of the terminal oxo ligand (figure 13). Cyclic voltammetry experiments were conducted to probe the redox chemistry of the two isomers. Both *trans* and *cis* isomers exhibit well-defined one-electron reductive signals for Mo^V → Mo^{IV} at

−940 mV and at −1160 mV respectively. Thus the results show that the *trans* isomer is easier to reduce than the *cis* isomer by about 200 mV provocatively suggesting that the position of the serinate oxygen relative to the terminal oxo ligand at the molybdenum center of DMSOR may play a critical role in gating the electron transfer process of active site regeneration as part of the catalytic cycle.[101] Another mono-oxo Mo^V complex, $Tp*Mo^VO(O-p-C_6H_4OEt)_2$, has been screened for its oxo-transfer ability. The electrochemical oxidation of this compound affords $[Tp*Mo^{VI}O(O-p-C_6H_4OEt)_2]^+$ which reacts with tertiary phosphines (PR_3) to generate a phosphineoxide-coordinated adduct, $[Tp*Mo^{IV}(OPR_3)(O-p-C_6H_4-OEt)_2]^+$. This compound subsequently eliminates OPR_3 to give the Mo^{IV} desoxo species, $[Tp*Mo^{IV}(O-p-C_6H_4OEt)_2]^+$. The desoxo species generates $Tp*Mo^VO(O-p-C_6H_4OEt)_2$ in the presence of water and an oxidizing agent completing the catalytic cycle, that closely resembles the postulated enzymatic turnover.[102]

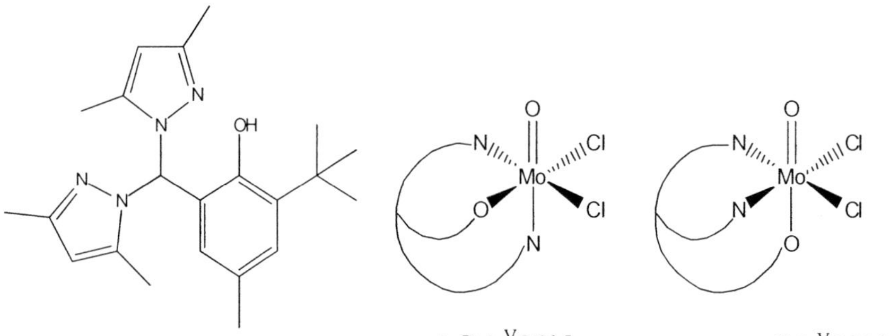

Figure 13. *cis* and *trans* orientation of the phenolato oxygen of L1OH with respect to the terminal oxo group in $MoOCl_2$.

The first oxo-bis-(dithiolene)metallate complexes to be characterized by X-ray crystallography were $[MoO(bdt)_2]^{2-}$ and $[MoO(bdt)_2]^-$ synthesized by the Garner group.[103] Following this, several bisdithiolene $Mo^{IV}O$ complexes were synthesized with symmetrical dithiolenes like $bdtCl_2$, mnt ($S_2C_2(CN)_2$), edt ($S_2C_2H_2$), $S_2C_2Ph_2$, $S_2C_2Me_2$ etc. and unsymmetrical dithiolenes like sdt, 2-pedt, 4-pedt (figure 14).[104-112] The preparation procedures for the bis(dithiolene)Mo(IV)O complexes are versatile and for a majority of these compounds summarized elsewhere in the literature.[7]

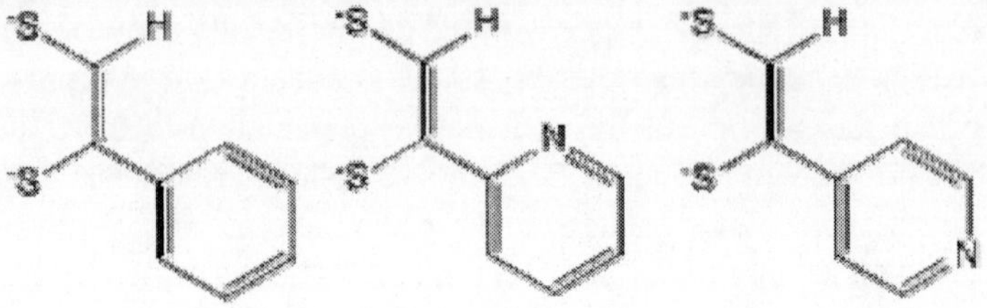

Figure 14. The unsymmetrical dithiolene ligands as models for MPT.

Bis-(dithiolene)mono-oxo Mo^{IV} complexes can be converted to the corresponding dioxo Mo^{VI} complexes through controlled oxidation by Me_3NO making model compounds for both the reduced and the oxidized oxidation state of the enzymes, and especially for AO, easily accessible at least when using plain dithiolene ligands. In recent years, the trend has been diverted from the above mentioned conventional dithiolene ligands to those with a closer resemblance of the molybdopterin structure. In 2005 $[MoO(fdt)_2]^{2-}$ (figure 15) was reported as the first molybdenum complex with a ligand system that not only includes the dithiolene group but also the pyrane feature of the enzymatic molybdopterin.[6] Crystals of these complexes could not be obtained possibly due to the fact that they form *cis*- and *trans*- (ligand orientation) and *R*- and *S*- (position of the phenyl group at the pyran ring) isomers *i.e.* a mixture of diastereomers, which usually results in poor crystallization behavior. Later in the same year, Sugimoto et al. synthesized the complexes $[MoO(L^{CH2})_2]^{2-}$, $[MoO(L^S)_2]^{2-}$ and $[MoO(L^O)_2]^{2-}$ (figure 15) with the latter again including a pyrane dithiolene molybdenum moiety. The crystal structures of these complexes show that the oxo ligand together with the four sulfur atoms from the dithiolene function constitute a square pyramidal geometry with a weakened Mo=O bond character.[113]

As mentioned earlier the $Mo^{IV}O(dt)_2/Mo^{VI}O_2(dt)_2$ (dt = dithiolene) couple was extensively utilized for modeling because of the frequent occurrence of this redox couple in conventional molybdenum chemistry. Even though these complexes do resemble the active site structure of AO, for a detailed understanding of structure function relationships in the majority of enzymes in the DMSOR family $Mo^{IV}(dt)_2/Mo^{VI}O(dt)_2$ couples need to be developed and studied carefully. This type of complexes, however, was not known to inorganic chemists until 1998 when Donahue et al. described the reactions 4 and 5 affording $[Mo^{IV}(bdt)_2(OSi^tBuPh_2)]^-$ and $[Mo^{VI}O(bdt)_2(OSi^tBuPh_2)]^-$. [112]

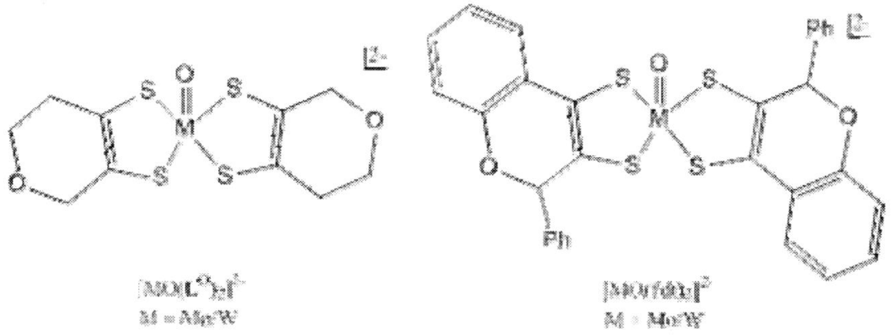

Figure 15. Molybdenum and tungsten complexes with pyrane dithiolene co-ordination.

$$[Mo^{IV}O(bdt)_2]^{2-} + Ph_2{}^{t}BuSiCl \rightarrow [Mo^{IV}(bdt)_2(OSi^{t}BuPh_2)]^{-} + Cl^{-} \quad (4)$$

$$[Mo^{VI}O_2(bdt)_2]^{2-} + Ph_2{}^{t}BuSiCl \rightarrow [Mo^{VI}O(bdt)_2(OSi^{t}BuPh_2)]^{-} + Cl^{-} \quad (5)$$

This approach was refined even further by the same group with respect to the modeling characteristics of the products by inclusion of alcoholate, thiolate and selenolate functions (reactions 6 and 7) giving compounds of the formula $[Mo(S_2C_2R_2)_2L(CO)_{2-n}]^{-}$ (n = 1 or 2; R= Me, Ph; L = $^{-}$OR, $^{-}$SR, $^{-}$SeR).[114]

$$[M(CO)_3(MeCN)_3] + 2[Ni(S_2C_2R_2)_2] \rightarrow$$
$$[M(S_2C_2R_2)_2(CO)_2] + [Ni^{II}{}_2(S_2C_2R_2)_2] + 3MeCN + CO \quad (6)$$

$$[M(S_2C_2R_2)_2(CO)_2] + L^{-} \rightarrow [M(S_2C_2R_2)_2L(CO)_{2-n}]^{-} + n\ CO \quad (7)$$

One or two labile carbonyl ligands in $[Mo(S_2C_2R_2)_2(CO)_2]$ are replaced by one mono-anionic ligand L^{-} (LH= PhOH, PhSH, PhSeH, 2,5,6-^{i}Pr$_3$PhOH, 2,5,6-^{i}Pr$_3$PhSH, 2,5,6-^{i}Pr$_3$PhSeH, AdOH, AdSH, AdSeH). The nature of the anionic ligand (sterical demand depending on ligand size and length of the metal donor bond) dictates the number of CO ligands that are leaving the complex. For instance, in the case of anionic oxygen ligands PhO^{-}, 2,5,6-^{i}Pr$_3$PhO^{-} and AdO^{-}, both carbonyl groups are replaced to afford complexes of the type $[M(S_2C_2R_2)_2L]^{-}$. PhS^{-} gives a monocarbonyl species $[M(S_2C_2R_2)_2L(CO)]^{-}$ but other bulkier thiolates replace both carbonyls leading to $[M(S_2C_2R_2)_2L]^{-}$. In case of selenolate the metal selenium bond is even longer and consequently PhSe^{-} and 2,5,6-^{i}Pr$_3$PhSe^{-} produce monocarbonyl species $[M(S_2C_2R_2)_2L(CO)]^{-}$ whereas the sufficiently bulkier AdSe^{-} replaces both carbonyls. The carbonyl displacing ability of the ligands are in the order: alcoholate > thiolate > selenolate, exactly opposed to the order in M-L bond length: M-O < M-S < M-Se. Thus, the mode of reactions described above is governed by both steric restrictions and the metal-ligand bond length. This may play a role in the enzymatic reactions as well. Similar reactions were carried out with other alcoholates like $^iPrO^-$ and $C_6F_5O^-$ to obtain $[M(S_2C_2R_2)_2L]^-$ complexes of square pyramidal geometry.[115] In order to synthesize an oxidized analogue to these Mo^{IV} complexes, $[Mo^{IV}(OPh)(S_2C_2Me_2)_2]^-$ was reacted with N-oxides and S-oxides to yield $[Mo^{VI}O(OPh)(S_2C_2Me_2)_2]^-$ in solution, but this species could not be isolated.

The enzyme selenate reductase is (as arsenite oxidase) another exception to the DMSOR family of enzymes. Based on EXAFS data its active site has been proposed to be of the composition $[Mo(MPT)_2OH]$ in the reduced and of $[MoO(MPT)_2OH]$ in the oxidized state.[84] The synthesis of active site analogues for this enzyme still remains a challenging task requiring the synthesis of bis(dithiolene) $Mo^{IV}(OH)$ and $Mo^{VI}O(OH)$ complexes. Wang et al. have synthesized a methoxy derivative of the above discussed molybdenum compounds $[Mo^{IV}(OMe)(S_2C_2Me_2)]^-$ by treating $[Mo(CO)_2(S_2C_2Me_2)_2]$ with NaOMe.[116] The methanolate ligand is comparably small and versatile because it can be regarded as model for both the hydroxide ligand of selenate reductase and the serinate residue in other enzymes of the DMSOR family with a molybdenum peptide bond. Interestingly it was found that the complex is both a structural model and a functional model cleanly reducing selenate to selenite in the reaction:

$$Mo^{IV}(OMe) + SeO_4^{2-} \rightarrow Mo^{VI}O(OMe) + SeO_3^{2-} \qquad (8)$$

Chapter 6

MODELING THE THREE FAMILIES OF TUNGSTEN ENZYMES

As stated previously tungsten enzymes fall into three major groups: the AOR, FDH and AH families (figures 2.l-n). The enzymes of the AOR family convert aldehydes to carboxylic acids.[2] The majority of tungsten enzymes belong to this family. The aldehye oxidoreductase from *Pyrococcus furiosus* (P^f AOR), a hyperthermophilic archaeon, is the first enzyme with a molybdopterin that has been structurally characterized by X-ray diffraction.[11] Analogous to the molybdenum enzymes of the DMSOR family two molybdopterin ligands were found to bind to tungsten. The tungsten atom and the two pairs of dithiolene sulfurs are arranged in a distorted square pyramid, with an angle between the planes of the molybdopterin ligands of ca. 97°. The two MPTs do not only bind to the tungsten but are also linked together through their phosphate functions, which coordinate axial sites of the same magnesium ion (figure 1.c). No coordinating protein ligands were found at tungsten, although electron density studies indicated the presence of two additional coordination sites at the W center. Based on the then available knowledge, Chan et al. proposed an additional occupation of these coordination sites by either glycerol or oxo ligands (or both) in a distorted trigonal prismatic geometry. It was later suggested that the observed glycerol stems from a protein storage buffer and may represent a substrate analogue.[117] Although there is considerable ambiguity about the additional coordination sites, it is likely that the oxidized enzyme is of $(MPT)_2W^{VI}O(OH)$ and the reduced enzyme of $(MPT)_2W^{IV}(OH)$ composition.[2] An EXAFS study by George et al. on *P. furiosus* AOR indicated the presence of an oxo group at 1.7 Å coordinating to the tungsten atom and an additional O or N atom possibly present at 2.1 Å.[118] Another crystallographically characterized enzyme of this family is *P. furiosus* formaldehyde ferredoxin oxidoreductase (Pf FOR).[119] As in the case of AOR, the tungsten atom is coordinated by four dithiolene sulfur atoms from two molybdopterins with an average W-S distance of 2.49 Å. There is no protein side-chain coordination to the tungsten atom and the two pterin molecules are linked to each other by a magnesium ion. Besides the four sulfur atoms, an additional ligand was found to bind to tungsten in Pf FOR, and it is assumed to be an oxygen atom. The potential difficulty in identifying additional coordination sites on the very heavy tungsten is supposed to be due to the heterogeneous nature (mainly the simultaneous occurrence of different oxidation states) of the tungsten site itself.

The second family of tungsten enzymes is the formate dehydrogenase (FDH) family. The two prominent members of this family are formate dehydrogenase (catalyzes conversion of formate to CO_2) and N-formylmethanofuran dehydrogenase (catalyzes conversion of N-formylmethanofuran to CO_2). For both types of enzymes the catalyzed reactions are reversible, which makes them interesting with respect to nowadays' greenhouse gas problem. The crystal structure of *Desulfovibrio gigas* formate dehydrogenase has been measured to a resolution of 1.8 Å.[120] In this enzyme tungsten is coordinated by the four dithiolate sulfur atoms from two MGDs, by the selenium atom of a SeCys and by one hydroxyl or sulfide ligand. Although X-ray absorption spectroscopy of the similar molybdenum site of *Escherichia coli* formate dehydrogenase seems to favor the OH ligand rather than sulfur[121] this observation is not transferable to tungsten FDH. The structural data by Raajimakers et al. favor a sulfur atom for the sixth ligand, although the resolution of the data is not sufficient to unambiguously distinguish between O and S. Interestingly, in a recent re-evaluation of the crystallographic data of the molybdenum-containing *E. coli* formate dehydrogenase originally recorded by Boyington et al.[122], Romão and co-workers proposed that the apical ligand was better refined as a sulfur atom (=S or –SH, not H_2O or –OH) at the molybdenum site.[123] And this would actually be analogous to the W-FDH from *D. gigas*. Although the available resolution does not allow an ultimate conclusion based on structural data regarding O and S, mechanistic evidence for the presence of a sulfur ligand has already been found in the inactivation of formate dehydrogenase from *Methanobacterium formicicum* using cyanide[124]: It was observed that incubation of the oxidized form of formate dehydrogenase with cyanide resulted in the release of equimolar amounts of thiocyanate and the subsequent deactivation of the enzyme. These observations in combination strongly suggest a sulfur ligand (most probably –SH) to be the sixth ligand in W-FDH.[15]

Acetylene hydratase is different from the other two families of tungsten enzymes and MPT dependent enzymes in general as it does not catalyze a redox reaction. Tungsten is not even believed to change its oxidation state during catalytic turn over. Instead it catalyzes the conversion of acetylene to acetaldehyde which is the addition of water to C_2H_2. Recently Seiffert et al. published the crystal structure of acetylene hydratase from *P. acetylenicus*.[17] The structure shows that the tungsten atom is coordinated by four sulfur atoms from the dithiolene groups of two MGD ligands and a fifth sulfur of a cysteine residue. A water molecule completes the slightly distorted octahedral geometry of the tungsten site.

Active site structures of several tungsten enzymes could not be determined yet due to crystallization difficulties and in addition the exact site structures of some of the already measured tungsten enzymes are not completely beyond doubt. Many enzymes show resemblance to the molybdenum enzymes to some extent and even isoenzymes of molybdenum and tungsten are described in the literature. For example tungsten-DMSOR is an isoenzyme of *R. capsulatus* Mo-DMSOR which is obtained when organisms are cultured in a medium with much higher concentration of tungstate than molybdate.[7] Usually these enzymes, in which one metal is replaced by the other, loose activity and the structural information obtained from those modified forms is most likely not relevant for the natural relatives. Information extracted from molybdenum enzymes may help in understanding certain aspects of tungsten enzymes but great care is needed when considering analogies and differences. In comparison much more and more precise structural and functional data is available for the more abundant molybdenum enzymes. And this is reflected in the synthetic model chemistry for these enzymes. Compared to the molybdenum analogues, tungsten analogues are less well studied and rarer systematically synthesized. Quite often, a tungsten model lacks the structural features that are essential for a particular family and advanced synthetic methodologies are yet to be developed. Naturally a large number of tungsten analogues of structural models originally developed for molybdenum enzymes is available and such complexes are essential in understanding the fundamental differences in the bioinorganic chemistry of both elements. In view of the less developed tungsten model chemistry and the structural similarity of the respective enzymes, tungsten model compounds will be discussed together in this chapter and not under separate headings.

As in the case of molybdenum model chemistry the $W^{IV}O(\text{dithiolene})_2$ and $W^{VI}O_2(\text{dithiolene})_2$ complexes have been investigated because of their frequent occurrence in inorganic chemistry. In addition bis(dithiolate) and *non*-dithiolene complexes of tungsten relevant for the oxo transfer enzymes have been developed. Many of them have been screened in the net reactions 9 and 10 for their oxo transfer ability.

$$W^{IV}O + XO \leftrightarrows W^{VI}O_2 + X \tag{9}$$

$$W^{IV} + XO \leftrightarrows W^{VI}O + X \tag{10}$$

Many six-fold coordinated $W^{VI}O_2$ complexes have been reported and a lesser number of corresponding $W^{IV}O$ complexes. Among those are the tungsten(VI) dithiocarbamate complexes, $W^{VI}O_2(R_2dtc)_2$ (R=Et, Me, ^{n}Pr; dtc = dithiocabamate).[125] These compounds were originally prepared by the oxo transfer reaction between $W(CO)_2(PPh_3)(R_2dtc)_2$ and $Mo_2O_3(Et_2dtp)_4$. Other complexes of this class are $(NH_4)_2[W^{VI}O_2(O_2CC(S)PPh_2)_2]$[126] and $W^{VI}O_2(ssp)$ (ssp= 2-(salicylideneamino)benzenethiolate), a Schiff base complex[127]. The synthesis of monooxo tungsten(IV) analogues of these complexes was found to be almost impossible. But stable bis(dithiolene) $W^{IV}O$, $W^{V}O$ and $W^{VI}O_2$ complexes with the ligands benzenedithiolene (bdt) and maleonitriledithiolene (mnt), which are much better models for MPT than the dithiocarbamates or Schiff bases, have been reported.[128,129] $(PPh_4)_2[W^{IV}O(bdt)_2]$ and $(NEt_4)_2[W^{IV}O(bdt)_2]$ were synthesized by borohydride reduction of $(PPh_4)[W^{V}O(bdt)_2]$ or $(NEt_4)[W^{V}O(bdt)_2]$ which were obtained by a simple ligand exchange reaction between $[WO(SPh)_4]^-$ and two equivalents of bdt-H_2. These complexes readily undergo oxidation by triethylamine-N-oxide to give the corresponding dioxo tungsten(VI) complexes. $[Et_4N]_2[W^{IV}O(mnt)_2]$ was prepared by the oxo transfer between $[Et_4N]_2[W^{VI}O_2(mnt)_2]$ and PPh_3. Remarkably the mnt complex $[Et_4N]_2[W^{IV}O(mnt)_2]$ was shown to reduce CO_2/HCO_3^- (at pH 7.5) to yield $HCOO^-$ and $[Et_4N]_2[W^{VI}O_2(mnt)_2]$ mimicking tungsten-formate dehydrogenase (W-FDH) activity.[130] A similar redox couple involving Me_3NO as a substrate analogue and $S_2C_2Me_2$ as the MPT analogue has been reported by the Holm group.[131] More recently, other important tungsten compounds with promising ligand systems involving pyrane dithiolene species have been synthesized. $[W^{IV}O(fdt)2]^{2-}$ was the first in this series, but no crystal structure was reported (figure 15).[6] Later, Sugimoto et al. synthesized $[WO(pdt)_2]^{2-}$ and $[WO_2(pdt)_2]^{2-}$ as the first structurally characterized tungsten complexes with pyrane dithiolene ligands (figure 15; L^0 = pdt).[132]

A series of various bis(dithiolene)tungsten(IV,VI) complexes with benzenedithiolene were prepared by Holm and co-workers in a synthetic approach to more accurate models for the active sites of tungstoenzymes.[13] Using $[W^{IV}O(bdt)_2]^{2-}$ as precursor, silylation with Me_3SiCl, $^tBuMe_2SiCl$ and $^tBuPh_2SiCl$ transforms the oxo ligand to a silyloxide ligand affording $[W^{IV}(bdt)_2(OSiMe_3)]^{1-}$, $[W^{IV}(bdt)_2(OSi^tBuMe_2)]^{1-}$ and $[W^{IV}(bdt)_2(OSi^tBuPh_2)]^{1-}$ respectively. Oxidation of the desoxo complex, $[W^{IV}(bdt)_2(OSi^tBuPh_2)]^{1-}$ with Me_3NO gives the corresponding monooxo tungsten(VI) complex $[W^{VI}O(bdt)_2(OSi^tBuPh_2)]^{1-}$. Silylation of $[W^{VI}O_2(bdt)_2]^{2-}$ with $^tBuPh_2SiCl$ affords the same complex. Further silylation of $[W^{VI}O(bdt)_2(OSi^tBuPh_2)]^{1-}$ with Me_3SiCl leads to the formation of $[W^{VI}O(bdt)_2Cl]^{1-}$ from which the unstable species $[W^{VI}O(bdt)_2L]^{1-}$ (L = $^tBuO^-$, PhS^-) were generated in solution. Interestingly, the reaction of $[W^{VI}O(bdt)_2Cl]^{1-}$ with L' = $P(OEt)_3$ or ^tBuNC afforded the reduced products $[W^{IV}(bdt)_2L'_2]$. Complexes $[W^{IV}(bdt)_2(OSiMe_3)]^{1-}$, $[W^{IV}(bdt)_2(OSi^tBuMe_2)]^{1-}$ and $[W^{IV}(bdt)_2(OSi^tBuPh_2)]^{1-}$ are regarded as tungsten analogues of the molybdenum model complexes for the reduced DMSOR family enzymes. However, the elusive $[W^{VI}O(OH)(dt)_2]$ and $[W^{IV}(OH)(dt)_2]$ complexes would be still better mimics of the active sites of tungsten-AOR family enzymes considering the current knowledge of the coordination environment of these enzymes, which was not available at the time of preparation. Further development of this chemistry by the Holm group led to similar models with less unnatural mono-anionic ligands ($[W^{VI}O(L^-)(dt)_2]$ / $[W^{IV}(L^-)(dt)_2]$; LH = PhOH, PhSH, PhSeH, ^iPrOH).[114,115] The reactions follow the equations 6 and 7 (see above) affording $[W(S_2C_2R_2)_2L(CO)_{2-n}]^-$ complexes in analogy to the molybdenum chemistry. In case of molybdenum complexes, we considered the L^- ligands as the structural analogues of the amino acids present at the molybdenum center of the DMSOR family of enzymes and this applies also to cysteine and selenocysteine at the active sites of the tungsten FDH and AH families. In addition the coordination environment of the above mentioned complexes $[W^{VI}O(^-OR)(dt)_2]$/ $[W^{IV}(^-OR)(dt)_2]$ can be regarded as analogous to that in the AOR family of enzymes. More recently the synthesis of $[W^{IV}(OMe)(S_2C_2Me_2)]^-$ was reported.[116] This complex shows a remarkable similarity to the coordination environment of tungsten in the AOR family of enzymes with the anionic oxygen based ligand carrying only a small methyl substituent.

As discussed above, in order to model the FDH family of enzymes a sulfur ligand is required though it is not known with certainty if it is a sulfide or a thiolate that is bound to tungsten at the active site. Very recently, the Holm group was successful in synthesizing $(Et_4N)[W^{VI}S(SeAd)(S_2C_2Me_2)_2]$ as the first FDH family analogue with the coordination of a sulfide ligand to tungsten.[133] Unfortunately a crystal structure or a functional oxo transfer study of this complex has not been reported yet.

It is clear from the above discussion that the tungsten model chemistry is still not advanced enough to address all the structural and reactivity issues and remains to pose a challenge to bioinorganic chemists working in this field. A brief discussion of these challenges is provided in the last chapter.

Chapter 7

OAT AND PCET REACTIONS

As stated previously molybdenum and tungsten enzymes (except nitrogenase) usually catalyze reactions of the type given in equation 1 in the introduction, leading to the net transfer of an oxygen atom to or from substrate. This type of reaction is generally called oxygen atom transfer (OAT) reaction without carrying a mechanistic implication and it is associated with a proton coupled electron transfer (PCET) process. PCET combines electron and proton transfers between reactants and products. A detailed survey of PCET processes in general has been published recently.[8] Oxygen atom transfer reactions relevant for the molybdenum enzymes have been extensively reviewed by Holm in 1990, prior to the elucidation of the active site structures of the MPT bearing enzymes.[134] Based on new and increasingly detailed insight into the active site structures of more and more enzymes, the related oxygen atom transfer model chemistry has also been subjected to reevaluation. In this section, we will be reviewing important OAT/PCET reactions by some of the complexes described above.

One of the most extensively studied redox couples is $Mo^{IV}O/Mo^{VI}O_2$ as a 'natural choice' due to the high frequency of occurrence of these groups in conventional molybdenum chemistry.[7] Several model reactions utilizing this redox couple have been reported and most of them were originally proposed or considered as relevant for the DMSO reductase family of enzymes. For example, the reaction 11 (where t-BuL-NS = bis(4-tert-butylphenyl)- 2-pyridylmethanethiolate) (figure 16) was studied with a wide variety of substrates by the Holm group.[54,135]

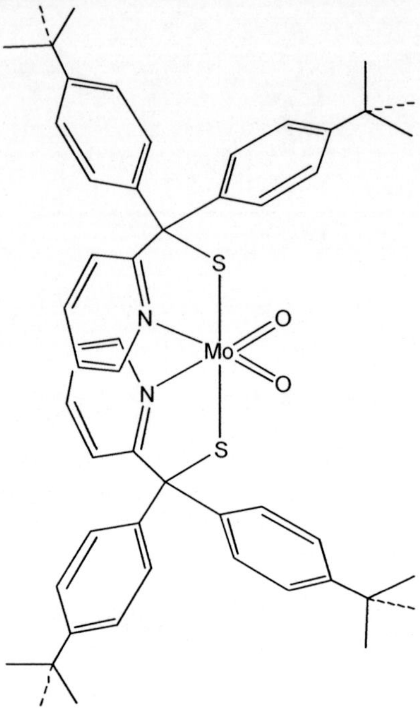

Figure 16. Chemical structure of MoO$_2$(t-BuL-NS)$_2$.

$$\text{MoO}_2(t\text{-BuL-NS})_2 + X \leftrightarrows \text{MoO}(t\text{-BuL-NS})_2 + XO \qquad (11)$$

Using isotope labeled substrates (^{18}O) oxygen atom transfer from substrate (XO = Ph$_2$S^{18}O) to complex and from complex to second substrate (X = Et$_3$P) was monitored in order to understand the relevant enzymatic reaction mechanisms. Later Holm and co-workers investigated OAT using the isolated enzyme *R. sphaeroides* DMSO reductase as catalyst and again isotope labeled substrate (Me$_2$S^{18}O) with a nonphysiological oxo acceptor, 1,3,5-Triaza-7-phosphatricyclo[3.3.1.1^{3,7}]decane (PTA) by absorption spectroscopy and CI mass spectrometry.[136] The results for the enzymatic experiments were very similar to those of the earlier model complex studies confirming their relevance. In the single turnover experiments anaerobic treatment of DMSO reductase with a large excess of the oxygen acceptor PTA resulted in replacement of the observed broad band in the UV-Vis spectrum at 720 nm with a less intense broad band at 640 nm over ca. 4 h corresponding to the reaction Enz Mo^{VI}O$_2$ + PTA → Enz Mo^{IV}O + PTAO. Further addition of Me$_2$SO led to a regeneration of the spectrum corresponding to the oxidized form of the enzyme. In reaction systems containing both Me$_2$S^{18}O and PTA in the presence of reduced enzyme ^{18}O isotope transfer from Me$_2$SO to PTA was detected by mass spectrometry. Since there was no reaction in the absence of enzyme, which remained intact during the reaction, it was concluded that the enzyme was a necessary mediator of the oxo transfer from Me$_2$SO to PTA by the following consecutive steps.

$$Me_2S(^{18}O) + Enz\ Mo^{IV}O \rightarrow Me_2S + Enz\ Mo^{VI}O(^{18}O) \qquad (12)$$

$$Enz\ Mo^{VI}O(^{18}O) + PTA \rightarrow Enz\ Mo^{IV}O + PTA(^{18}O) \qquad (13)$$

Naturally, similar reactions were studied for Mo^{IV}O/Mo^{VI}O$_2$ redox couples carrying dithiolene ligands. Mo^{IV}O(dithiolene)$_2$ type complexes can be oxidized to the corresponding Mo^{VI}O$_2$(dithiolene)$_2$ complexes upon reaction with strong oxidizing agents as Me$_3$NO (equation 14).[105] The backward transformation follows the reaction 15 and has been reported in the case of [Mo^{VI}O$_2$(mnt)$_2$]$^{2-}$ using oxo acceptors as Et$_3$P, PhEt$_2$P, Ph$_2$EtP and PPh$_3$.

$$[MoIVO(dt)_2]^{2-} + Me_3NO \rightarrow [Mo^{VI}O_2(dt)_2]^{2-} + Me_3N \qquad (14)$$

$$[Mo^{VI}O_2(dt)_2]^{2-} + R_3P \rightarrow [Mo^{IV}O(dt)_2]^{2-} + R_3PO \qquad (15)$$

However, it is well known that the formation of dimeric μ-oxo Mo^V species from a comproportionation of oxidized and reduced species often prevents Mo^{IV}O complexes from being effective OAT catalysts.[134,137] Successfully preventing this kind of deactivation by steric control of the ligands has already been mentioned when discussing the sulfite oxidase model chemistry. However, a second strategy emerged based on the reported μ-oxo Mo^V$_2$ complexes showing a *cis*-configuration between terminal and μ-oxo ligands.[138,139] This suggests a possible formation of the dimer by attack of one of the oxo ligands of the Mo^{VI}O$_2$ species to the vacant *trans* position of the Mo^{IV}O species with a subsequent *trans-cis* rearrangement between the dithiolene and oxo ligands. Nakamura and coworkers studied this property in detail and put forward the idea of prevention of deactivation due to dimerization by an electronic control of the ligands.[105,140,141] For example benzenedithiolene Mo^{IV}O complexes react with Me$_3$NO to afford the corresponding Mo^{VI}O$_2$ species but do not lead to μ-oxo Mo^V$_2$. This reflects the difficulty of ligand dissociation from the metal caused by the strongly chelating character of this ligand. At the same time the reaction between (NEt$_4$)$_2$[Mo^{IV}O(bdt)$_2$] and (NEt$_4$)$_2$[Mo^{VI}O$_2$(S$_2$CNEt$_2$)$_2$] (S$_2$CNEt$_2$ = N,N-diethyldithiocarbamato) in acetonitrile proceeds in an interesting and unanticipated way. The only products obtained were (NEt$_4$)[Mo^{V}O(bdt)$_2$] and [(Mo^{V}O)$_2$(μ-O)(S$_2$CNEt$_2$)$_2$] while the expected mixed μ-oxo Mo^V species with both ligand types was not detected. This indicates that the stronger binding of the bdt ligand compared to the S$_2$CNEt$_2$ ligand prevents any isomerization and comproportionation reaction. These results suggest that the reactivity of the Mo^{IV}O or Mo^{VI}O$_2$ species can be controlled not only sterically but also electronically by using strongly chelating dithiolene ligands in order to inhibit the formation of the (μ-oxo)-dimolybdenum(V) center.

Another aspect with respect to structure function relations is that of the different coordinated amino acids (Ser, Asp, Cys, Se-Cys) at the active sites of the DMSO reductase family enzymes. It is not known today if the different amino acids are important and specific for each enzyme or if this part of the active site was developed rather randomly. In one case though (periplasmatic nitrate reductases) the cysteine residue is proposed to actively take part in bond breaking and bond making at the substrate.[142,143]

In order to evaluate the influence of sulfur versus selenium coordination model compounds have been developed (without consideration of dithiolene coordination) in which sulfur and selenium occupy analogous functions and positions. Ma et al. reported the synthesis and structure of dinuclear-oxomolybdenum(V) complexes [Mo$_2$O$_3$(PyS)$_4$], [Mo$_2$O$_3$(PySe)$_4$] and [Mo$_2$O$_3$(4-CF$_3$-PymS)$_4$] by reactions of the [MoO$_2$Cl$_2$(DME)] precursor with the corresponding heterocyclic bidentate (N,X) ligands, X = S, Se, where PyS, PySe and 4-CF$_3$-PymS are the anions of pyridine-2-thione, pyridine-2-selenolato and 4-trifluoromethyl-2-pyrimidinthiol, respectively.[144] The oxo-transfer catalytic activities of [Mo$_2$O$_3$(PyS)$_4$] and [Mo$_2$O$_3$(PySe)$_4$] were studied in the reaction between PPh$_3$ and DMSO under analogous conditions. The reactions follow the equations 16 and 17.

$$Mo^V{}_2O_3L_4 + DMSO \rightarrow 2Mo^{VI}O_2L_2 + DMS \qquad (16)$$

$$2Mo^{VI}O_2L_2 + PPh_3 \rightarrow Mo^V{}_2O_3L_4 + OPPh_3 \qquad (17)$$

It was found that the reaction rate was approximately tenfold increased in the case of [Mo$_2$O$_3$(PyS)$_4$] compared to its selenium counterpart [Mo$_2$O$_3$(PySe)$_4$]. So the catalytic performance of both compounds is strongly influenced by an exchange of ligand atoms (sulfur versus selenium) even though no substantial influence on the structures could be observed. Another pair, consisting of the first crystallographically characterized molybdenum(VI) selenoether complex [Mo$_2$O$_4$(OC$_3$H$_6$SeC$_3$H$_6$O)$_2$] and its thioether analogue [Mo$_2$O$_4$(OC$_3$H$_6$SC$_3$H$_6$O)$_2$], was synthesized and investigated in comparison.[145] Both structures are almost identical except for those parameters that are directly derived from the different sizes of the varied ligand atoms (Se and S). The metal centered redox process ($Mo^V \leftrightarrow Mo^{VI}$) is at slightly lower voltage (0.045 V) for the sulfur containing compound than for the selenium containing compound (0.114 V). Nevertheless, the selenium compound catalyzes the oxygen atom transfer from DMSO to PPh_3 by a different mechanism and at a much higher rate than the sulfur compound (in contrast to the previous study). The dimeric complexes investigated in these two studies are no structural models for the enzyme's active sites but they were able to effectively catalyze oxygen atom transfer (100% conversion) without any occurrence of inactive complex side products and delivered results, which indicated that cysteine and selenocysteine might indeed be used for a purpose in the different molybdenum and tungsten cofactors.

The first example of a model complex with both structural and functional relevance especially for arsenite oxidase was reported by the Sugimoto group.[146] The respective $Mo^{IV}O/Mo^{VI}O_2$ redox couple was shown to be able to participate in both OAT and PCET reactions. The $(Bu_4N)_2[Mo^{IV}O(bdtCl_2)_2]$ complex ($bdtCl_2$ = 3,6-dichloro-1,2-benzenedithiolate) is converted to $(Bu_4N)_2[Mo^{VI}O_2(bdtCl_2)_2]$ in aqueous media by a PCET process: $(Bu_4N)_2[Mo^{IV}O(bdtCl_2)_2]$ undergoes chemical oxidation by two equivalents of $K_3[Fe(CN)_6]$ in the presence of two equivalents of ^tBuOK in CH_3CN/H_2O (5:7) yielding $(Bu_4N)_2[MoO_2(bdtCl_2)_2]$. When 98% $H_2^{18}O$ was used instead of H_2O, the product obtained was $(Bu_4N)_2[Mo^{VI}O^{18}O(bdtCl_2)_2]$ proving water to be the source of oxygen. In the second step $(Bu_4N)_2[MoO_2(bdtCl_2)_2]$ transfers this oxygen atom to arsenite forming arsenate and giving back the mono oxo species $(Bu_4N)_2[Mo^{IV}O(bdtCl_2)_2]$ (figure 17).

However, the solution of an increasing number of enzymatic active site structures led to the necessity to utilize the chemically much more elusive $Mo^{IV}/Mo^{VI}O$ redox couple for OAT reactions with respect to the DMSOR family. The synthetic methods developed by the Holm group leading to this kind of complexes have already been discussed in detail in chapter 5. $[Mo^{IV}(OPh)(S_2C_2Me_2)_2]^-$ reacts with XO = Me_3NO and Me_2SO as oxygen donors to give $[Mo^{VI}O(OPh)(S_2C_2Me_2)_2]^-$ according to reaction 18.[115]

$$[Mo^{IV}(OPh)(S_2C_2Me_2)_2]^- + XO \rightarrow [Mo^{VI}O(OPh)(S_2C_2Me_2)_2]^- + X \qquad (18)$$

The oxo transfer in equation 18 has again been proven by isotope (^{18}O) labeling. The reduction of Me_2SO to Me_2S and of Me_3NO to Me_3N in acetonitrile at 298 K follows second order kinetics as shown by UV-Vis studies. The oxidized species

$[Mo^{VI}O(OPh)(S_2C_2Me_2)_2]^-$ is too unstable to be isolated and decays by an internal redox process to a Mo^VO product, liberating an equimolar amount of phenol as per reaction 19.

$$[Mo^{VI}O(OPh)(S_2C_2Me_2)_2]^- \rightarrow [Mo^VO(S_2C_2Me_2)_2]^- + PhO^\bullet \qquad (19)$$

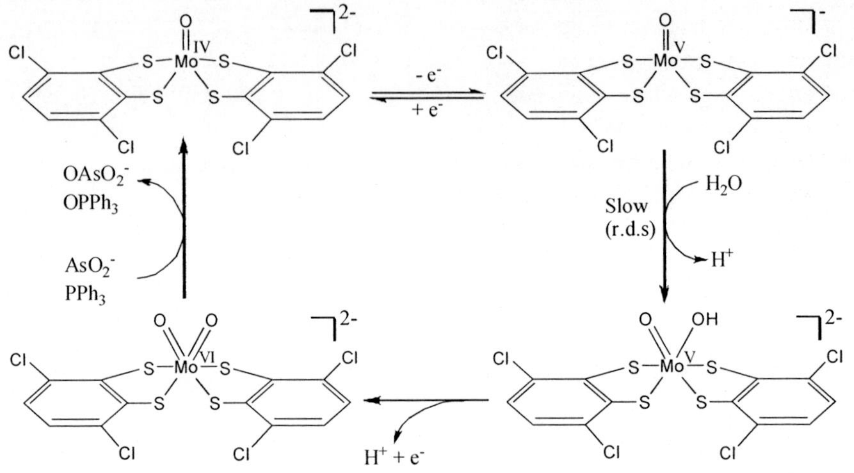

Figure 17. Proposed catalytic cycle involving the $Mo^{IV}O/Mo^{VI}O_2$ redox couple, participating in both PCET and OAT steps.

The reaction system in 18 is considered to be the first relevant functional model system for oxygen atom transfer with respect to the DMSO reductase family of enzymes. So the reaction schemes represented by equation 12 and 13, which were once considered to be relevant for the DMSOR enzymes, were rewritten to equations 20 and 21 to comply with the new structural information.[7]

$$Me_2S(^{18}O) + Enz\ Mo(IV) \rightarrow Me_2S + Enz\ Mo(VI)(^{18}O) \qquad (20)$$

$$Enz\ Mo(VI)(^{18}O) + PTA \rightarrow Enz\ Mo(IV) + PTA(^{18}O) \qquad (21)$$

A reaction cycle between desoxo Mo^{IV} and $Mo^{VI}O$ centers including both PCET and OAT was first reported by Kirk, Basu and coworkers (see above).[102] One electron oxidation of $LMo^VO(p\text{-}OC_6H_4\text{-}OC_2H_5)_2$ by $(NH_4)_2Ce(NO_3)_6$ leads to the formation of the corresponding mono oxo Mo^{VI} species $[LMo^{VI}O(p\text{-}OC_6H_4\text{-}OC_2H_5)_2]^+$ with L^- = hydrotris(3,5-dimethyl-1-pyrazolyl)borate. Acetonitrile solutions of cationic $[LMo^{VI}O(p\text{-}OC_6H_4\text{-}OC_2H_5)_2]^+$ undergo OAT to yield the cationic desoxo species [LMo^{IV}(p-OC$_6$H$_4$–OC$_2$H$_5$)$_2$]$^+$ and OPPh$_3$ when treated with PPh$_3$. In presence of water and the oxidizing agent 2,3-dicyano-5,6-dichloro-1,4-benzoquinone (DDQ) the desoxo species is transformed back into the LMo^{V}O(p-OC$_6$H$_4$–OC$_2$H$_5$)$_2$ complex completing the catalytic cycle (figure 18). When H$_2$^{18}O was used in the final step LMo^V(^{18}O)(p-OC$_6$H$_4$–OC$_2$H$_5$)$_2$ was formed confirming that water acts as the source of oxygen in the entire catalysis.

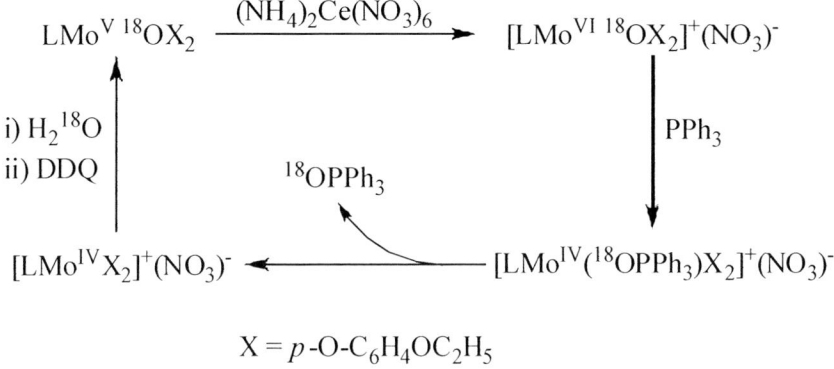

Figure 18. The reaction cycle combining PCET and OAT between desoxo Mo^{IV} and Mo^{VI}O centers.

The W^{IV}O/W^{VI}O$_2$ and W^{IV}/W^{VI}O redox couples are also well known by now but not as intensely studied as their molybdenum counterparts. For example [W^{IV}O(bdt)$_2$]$^{2-}$ and [W^{IV}O(S$_2$C$_2$Me$_2$)$_2$]$^{2-}$ complexes are readily oxidized by Me$_3$NO yielding the corresponding dioxo W^{VI} species [W^{VI}O$_2$(bdt)$_2$]$^{2-}$ and [W^{VI}O$_2$(S$_2$C$_2$Me$_2$)$_2$]$^{2-}$ respectively.[128] The oxo transfer ability of tungsten complexes utilizing the W^{IV}/W^{VI}O couple was studied for [W(OSiPh$_2$'Bu)(bdt)$_2$]$^{2-}$ with various substrates as Me$_3$NO, N-morpholine-N-oxide, DMSO, Ph$_3$PO, Ph$_2$SeO and Ph$_3$AsO.[13] The observed reactions were comparatively slow except in the case of Me$_3$NO and N-morpholine-N-oxide presumably due to the bulkiness of the ($^-$OSiPh$_2$'Bu) ligand. As in the case of molybdenum complexes, [W^{IV}(OPh)(S$_2$C$_2$R$_2$)$_2$]$^{2-}$ undergoes oxo transfer with XO (= Me$_3$NO, DMSO and Ph$_3$AsO) leading to the corresponding mono-oxo W^{VI} species [W^{VI}O(OPh)(S$_2$C$_2$R$_2$)$_2$]$^{2-}$.[131,147-149] Notably, the axial ligands were found to have a profound influence on the activity of the complex due to an electronic fine tuning of the central metal. If the axial ligand is iso-propoxide for instance, the activity diminishes to a rate which is 200 times slower than that of the phenoxide analogue for the same substrate.

Chapter 8

FUTURE PERSPECTIVES

The previous discussion has shown that structural and functional molybdenum and tungsten cofactor model chemistry is one of the hottest topics in bioinorganic chemistry due to its comprehensive importance in all domains of life. The synthetic challenges lying ahead are put forward by structural biologists and spectroscopists and often these challenges appear to be chemically unaccomplishable despite the existing natural composition of the active sites. One of the most interesting features of the molybdenum and tungsten enzymes (except nitrogenase) is the metal carrying the common and unique ligand molybdopterin. Since MPT binds to the metal by a dithiolene function, the respective *non-innocent* nature of this ligand plays a significant role in the redox properties of the active site. As a consequence the redox properties of the enzyme's active sites cannot be regarded as strictly metal based. Effectively the metal is part of an electronically delocalized system constituted by the planar MS_2C_2 ring. This electron delocalization allows the dithiolene function to be a mediator for the electronic coupling between the metal center and the pterin nucleus.[19] The coupling will be highly favored if the carbon atom adjacent to the dithiolene group is sp^2 hybridized because this allows conjugation between the metal dithiolene system and the pyrazine ring.[150] The entire role of the pterin and pyrane rings in the MPT structure has not been clarified yet even though its participation in fine tuning the redox nature of the active site is beyond doubt. It has to be mentioned here that the bicyclic MPT structure originally proposed by Rajagopalan and the tricyclic form found in X-ray crystal structures of MPTs in the following years is different only by two hydrogens. At the same time, the X-ray crystal structure of molybdenum nitrate reductase showed the presence of two MPTs at the metal center, one with a bicyclic open alcohol structure and the other with a tricyclic closed pyrane ring structure.[151] This indicates the possibility of pyrane ring opening and closing during the catalytic turnover effecting a redox action in the adjacent pterin ring which is conjugated to the metal-dithiolene aromatic system. The pyarazine ring present in the tricyclic form of MPT is in its reduced tetrahydro form as evidenced by crystallographic structures, but the opening of the pyrane ring may result in a dihydropyrazine structure and therefore the loss of two protons coupled to an oxidation of this ring. The reversal of this process is a two proton reduction (addition of protons) combined with the pyrane ring closing and restoration of the reduced tetrahydro pyrazine. A scheme illustrating the opening and closing of the pyrane in the MPT structure is shown in figure 19.[18] Consequently MPT is believed to have a considerable role in handling the two protons involved in the catalytic PCET reaction.

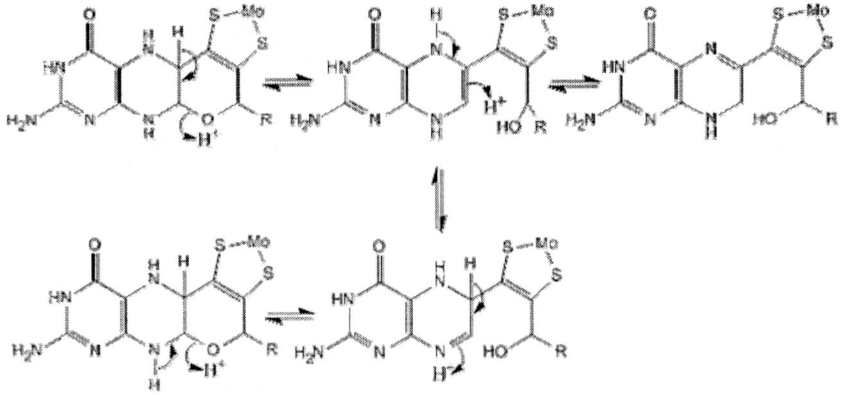

Figure 19. Illustration of the opening and closing of the pyrane in the MPT structure.

The bioinorganic chemist's biggest challenge, in our opinion, is to synthesize molybdenum or tungsten dithiolene complexes involving both the pyrane and pyrazine rings in the same ligand system. Such a synthetic MPT analogue remains elusive still. Pyrane dithiolene complexes of molybdenum and tungsten are known and have been discussed above. However, these complexes are unable to model the proposed ring opening and closing of the pyrane due to the absence of the neighboring pyrazine ring. Shown in figure 20 are the auspicious analogous CpCo(III) (where Cp = η^5-cyclopentadienyl) complexes involving dithiolene coordination as part of a dihydropyran fused to a pyrazine ring, which were prepared in Joule's laboratory.[152,153] Unfortunately no molybdenum or tungsten complexes have been prepared with these extremely promising ligands at least not with a yield allowing characterization.

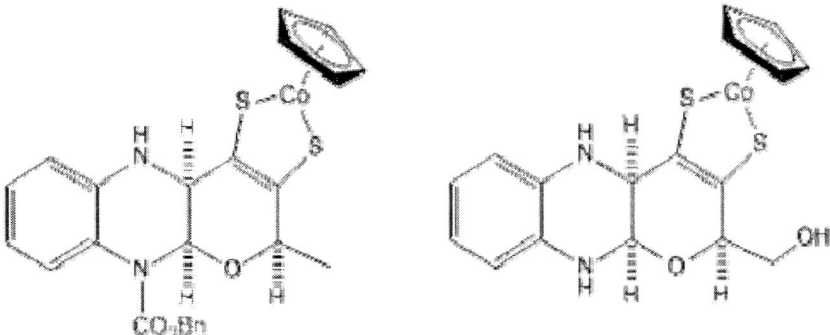

Figure 20. Co(III) complexes involving dithiolene coordination as part of a dihydropyran fused to a pyrazine ring.

Very recently Basu and co-workers were successful in developing a related ligand system, 4,4-Dimethyl-4H-5-oxa-1,3-dithia-6,11-diaza-cyclopenta[a]anthracen-2-one, in the completely oxidized form but with a closed pyrane ring, which is somehow a mixed model for the proposed pyrane activity [154]. The compound was used as new fluorescent sensor that can detect Pb^{2+} in aqueous solution over a wide pH range of 4 to 10 and in a mixture of several other metals at a concentration as low as 10 p.p.b.. For this reason, this new organic compound is called lead glow the structure of which is given in figure 21. However coordination of this compound to Mo or W has not been reported yet.

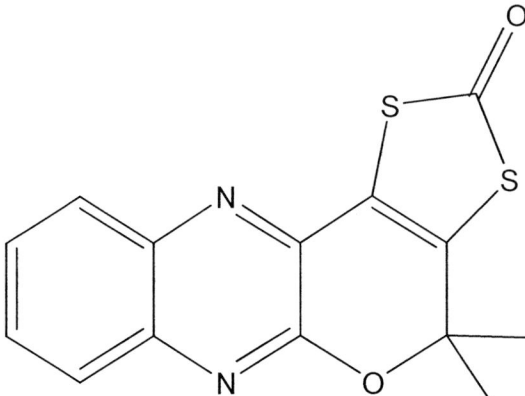

Figure 21. A ligand system in which a pyran ring is fused to the fully oxidized pyrazine ring.

A complete MPT model complex of molybdenum or tungsten remains to be one of the biggest challenges in bioinorganic chemistry requiring both inorganic and organic synthetic expertise, a close and inspiring cooperation between disciplines and very persistent experimenters.

Based on the crystallographic clarification of various enzymatic active sites' structures, we have at present a more or less clear focus on the synthetic analogues requiring preparation. In the past, bioinorganic chemists were successful to a great extent in this respect, but several unusual coordination features of the active sites are still not known in synthetic chemistry. In addition the oxo transfer abilities of many already developed model complexes are not fully explored with respect to the most relevant reactions. The most important and still elusive coordination features in the focus of bioinorganic chemists working in this field include bis(dithiolene) complexes of the $Mo^{VI}O(OH)/Mo^{IV}(OH)$ couple relevant for the selenate reductase active site, monodithiolene complexes of the $Mo^{VI}O(S)X/Mo^{IV}O(SH)X$ couple(X= OH, H_2O or N) relevant for the xanthine oxidase family of enzymes, bis(dithiolene) complexes of $Mo^{VI}S(SeS)$ relevant for the Mo-FDH active site, monodithiolene complexes of the $Mo^{VI}O(OH)(\mu\text{-}S)Cu^{I}/Mo^{IV}O(OH_2)(\mu\text{-}S)Cu^{I}$ couple relevant for CODH and finally bis(dithiolene) complexes of $W^{VI}(OH_2)(SR)$ relevant for the W-AH active site. $(Et_4N)[W^{VI}S(SeAd)(S_2C_2Me_2)]$ has already been synthesized as a structural analogue of the active site of FMDH, but a crystal structure could not be obtained and therefore the important structural information remains elusive.[133]

Several still not unanimously identified functional groups at the active sites of molybdenum and tungsten enzymes, new structures of so far unknown active sites and revision of old structures continue to challenge bioinorganic chemists' creativity, imagination and enthusiasm and promise to keep this field at the very frontier of bioinorganic research for quite a while.

ABBREVIATIONS

Ad	Adamantyl
AH	Acetylene hydratase
AO	Arsenite oxidase
AOR	Aldehyde oxido reductase
Ar	Aryl
Asp	Aspartate
bdt	Benzenedithiolate
bdtCl$_2$	3,6-Dichlorobenzene-1,2-dithiol
Cys	Cysteine
DDQ	2,3-Dicyano-5,6-dichloro-1,4-benzoquinone
DME	Ethyleneglycol dimethylther
DMSO	Dimethylsulfoxide
dtc	Dithiocarbamate
dtp	Dithiophosphinate
E. coli	Escherichia coli
edt	Ethene-1,2-dithiolate
EXAFS	Extended X-ray absorption fine structure
FDH	Formate dehydrogenase
fdt	Flavano-ene-dithiolate
FOR	Formaldehyde ferredoxin oxidoreductase
^{i}Pr	*iso*-Propyl
Me	Methyl
MGD	Molybdopterin guanine dinucliotide
mnt	2,3-Dimercaptomaleonitrile (maleonitriledithiolene)
MPT	Molybdopterin
OAT	Oxygen atom transfer

p.p.b.	Parts per billion
p.p.m.	Parts per million
PCET	Proton coupled electron transfer
pdt	Pyrane-ene-dithiolate
Ph	Phenyl
PTA	1,3,5-Triaza-7-phosphatricyclo[3.3.1.1^{3,7}]decane
PymS	Pyrimidinthiol
PyS	Pyridine-2-thionate
PySe	Pyridine-2-selenolato
r.d.s.	Rate determining step
SECIS	Selenocysteine insertion sequence
Se-Cys	Selenocysteine
Ser	Serine
ssp	2-(Salicylideneamino)benzenethiolate)
^{t}Bu	*tert*-Butyl
TMAOR	Trimethylamine *N*-oxide reductase
Tp*	Hydrotris-(3,5-dimethylpyrazol-1-yl)borate

REFERENCES

[1] Kletzin, A.; Adams, M. W. W. *FEMS Microbiol. Rev.* 1996, 18, 5-63
[2] Hille, R. *Trend. Biochem. Sci.* 2002, 27, 360-367
[3] Scott, C.; Lyons, T. W.; Bekker, A.; Shen, Y.; Poulton, S. W.; Chu, X.; Anbar, A. D. *Nature* 2008, 452, 456-460
[4] Bertine, K. K.; Turekian, K. K. *Geochim. Cosmochim. Acta* 1973, 37, 1415-1434
[5] Taylor, S. R.; McLennan, S. M. Rev. *Geophys.* 1995, 33, 241-265
[6] Schulzke, C. *Dalton Trans.* 2005, 713-720
[7] Enemark, J. H.; Cooney, J. J. A.; Wang, J.J.; Holm, R. H. *Chem. Rev.* 2004, 104, 1175-1200
[8] Huynh, M. H. V.; Meyer T. J. *Chem. Rev.* 2007, 107, 5004-5064
[9] Johnson, J. L.; Rajagopalan, K. V. *Proc. Nat. Acad. Sci. USA.* 1982, 79, 6856-6860
[10] Kramer, S. P.; Johnson, J. L.; Ribeiro, A. A.; Millington, D. S.; Rajagopalan, K. V. *J. Biol. Chem.* 1987, 262, 16357-16363
[11] Chan, M. K.; Mukund, S.; Kletzin, A.; Adams, M. W. W.; Rees, D. C. *Science* 1995, 267, 1463-1469
[12] Einsle, O.; Tezcan, F.A.; Andrade, S. L .A.; Schmid, B.; Yoshida, M.; Howard, J.B.; Rees, D.C. *Science* 2002, 297, 1696-1700
[13] Lorber, C.; Donahue, J. P.; Goddard, C. A.; Nordlander, E.; Holm, R. H. *J. Am. Chem. Soc.* 1998, 120, 8102-8112
[14] Stiefel, E. I. *Dalton Trans.* 1997, 3915-3923
[15] Bevers, L. E.; Hagedoorn P.-L., Hagen, W. R. *Coord. Chem. Rev.* 2009, 253, 269-290
[16] Hille R. *Chem. Rev.* 1996, 96, 2757-2816

[17] Seiffert, G. B.; Ullmann, G. M; Messerschmidt, A.; Schink, B.; Kroneck, P. M. H.; Einsle O. *Proc. Nat. Acad. Sci. USA* 2007, 104, 3073-3077
[18] Sugimoto, H; Tsukube, H. *Chem. Soc. Rev.* 2008, 37, 2609-2619
[19] Enemark, J.H.; Garner, C.D. *J .Bio. Inorg. Chem.* 1997, 2, 817-822
[20] George, G.N. *J. Bio. Inog. Chem.* 1997, 2, 790-796
[21] Mueller-Westerhoff, U.T.; Vance, B. In *Comprehensive Coordination Chemistry*, Wilkinson, G.; Gillard, R.D.; McCleverty, J.A.; Ed.; Pergamon: Oxford, 1987
[22] McCleverty, J.A. *Prog. Inorg. Chem.* 1968, 10, 49-221
[23] Eisenberg, R.; Ibers, J. A. *Inorg. Chem.* 1965, 4, 605-608
[24] Schrauzer, G. N.; Mayweg, V. *J. Am. Chem. Soc.* 1962, 84, 3221-3221
[25] Sartain, D.; Truter, M. R. *J. Chem. Soc. A* 1967, 1264-1272
[26] Eisenberg, R.; Ibers, J. A. *J. Am. Chem. Soc.* 1965, 87, 3776-3778
[27] Eisenberg, R.;Ibers, J.A. *Inorg. Chem* 1966, 5, 411-416
[28] Smith, A. E.; Schrauzer, G. N.; Mayweg, V. P.; Heinrich, W. *J. Am. Chem. Soc.* 1965, 87, 5798-5799
[29] Eisenberg, R.; Gray, H. B. *Inorg. Chem.* 1967, 6, 1844-1849
[30] Beswick, C. L.; Schulman, J.M.; Stiefel, E.I. In Dithiolene chemistry: synthesis properties and applications; Stiefel, E. I.; Ed.; *Progress in Inorganic chemistry*, Vol. 52, John Wiley & sons, Inc. Hoboken, New Jersey, 2004, Vol. 52, 55-110
[31] Alvarez, S.; Vicente, R.; Hoffmann, R. *J. Am. Chem. Soc* 1985, 107, 6253-6277
[32] Stiefel, E. I.; Eisenberg, R.; Rosenberg, C.; Gray, H. B. *J. Am. Chem. Soc.* 1966, 88, 2956-2966
[33] Eisenberg, R. ; Stiefel, E. I.; Rosenberg, R. C.; Gray, H. B. *J. Am. Chem. Soc.* 1966, 88, 2874-2876
[34] Schlaepfer, C. W.; Nakamoto, K. *Inorg. Chem* 1975, 14, 1338-1344
[35] Clark, R. J. H.; Turtle, P. C. *Dalton Trans* 1977, 2142-2148
[36] Schrauzer, G.N. *Acc. Chem. Res.* 1969, 2, 72-80
[37] Johnson, J.; Rajagopalan, K.V. *J Biol. Chem.* 1977, 252, 2017-2025
[38] Kipke, C. A., Cusanovich, M.A.; Tollin, G. ; Sunde, R. A.; Enemark, J.A. *Biochemistry* 1988, 27, 2918-2926
[39] Sullivan Jr., E. P.; Hazzard, J. T.; Tollin, G.; Enemark, J.H. *Biochemistry* 1993, 32, 12465-12470
[40] Rajagopalan, K. V. In *Molybdenum and Molybdenum Containing Enzymes*; Coughlan, M.; Ed.; Pergamon: Oxford 1980, 241-272
[41] Cramer, S. P.; Wahl, R.; Rajagopalan, K.V. *J. Am. Chem. Soc.* 1981, 103, 7721-7727

[42] Cramer, S. P.; Solomonson, L. P.; Adams, M. W. W.; Mortenson, L. E. *J. Am. Chem. Soc.* 1984, 106, 1467-1471
[43] George, G. N.; Kipke, C.A.; Prince, R. C.; Sunde, R. A.; Enemark, J. H.; Cramer, S. P. *Biochemistry* 1989, 28, 5075-5080
[44] George, G. N.; Garrett, R. M., Prince, R.C.; Rajagopalan, K.V. *J. Am. Chem. Soc.* 1996, 118, 8588-8592
[45] Kisker, C.; Schindelin, H.; Pacheco, A.; Wehbi, W. A.; Garrett, R.M.; Rajagopalan, K.V.; Enemark, J. H.; Rees, D.C. *Cell* 1997, 91, 973-983
[46] Garton, S. D.; Garrett, R.M.; Rajagopalan, K.V.; Johnson, M.K. *J. Am. Chem. Soc.* 1997, 119, 2590-2591
[47] George, G. N.; Pickering, I. J; Kisker, C. *Inorg. Chem.* 1999, 38, 2539-2540
[48] Hille, R. *J. Biol. Inorg. Chem.* 1997, 2, 804-809
[49] Fischer, B.; Enemark, J. H.; Basu, P. *J. Inorg. Biochem.* 1998, 72, 13-21
[50] Berg, J. M.; Holm, R. H. *J. Am. Chem. Soc.* 1985, 107, 917-925
[51] Berg, J. M.; Holm, R. H. *J. Am. Chem. Soc.* 1985, 107, 925-932
[52] Palanca, P.; Picher, T.; Sanz, V., Romero, P.G.; Llopis, E.; Domenech, A.; Cervilla, A., *Chem. Commun.* 1990, 531-533
[53] Gheller, S. F., Schultz, B. E.; Scott, M. J.; Holm, R. H. *J. Am. Chem. Soc.* 1992, 114, 6934-6935
[54] Schultz, B. E., Gheller, S. F.; Muetterties, M. C.; Scott, M. J.; Holm, R. H.; *J. Am. Chem. Soc.* 1993, 115, 2714-2722
[55] Bray, R.C. *Rev. Biophys.* 1988, 21, 299-301
[56] Xiao, Z.; Young, C. G.; Enemark, J. H.; Weddt, A. G., *J. Am. Chem. Soc.* 1992,114, 9194-9195
[57] Xiao, Z.; Bruck, M. A.; Enemark, J. H.; Young,C. G.; Wedd, A. G. *Inorg. Chem.* 1996, 35, 7508-7515
[58] Mader, M. L.; Carducci, M. D.; Enemark, J. H. *Inorg. Chem.* 2000, 39, 525-531
[59] Lim, B. S.; Willer, M. W.; Miao, M. Holm, R. H. *J. Am. Chem. Soc.* 2001, 123, 8343-8349
[60] Bray, B .J.; Garner, R. C.; Gutteridge, C. D.; Hasnain, S. S. *J. Biochem* 1980, 191, 499-508
[61] Cramer, S. P., Hille, R. *J. Am. Chem. Soc.* 1985,107, 8164-8169
[62] Turner, N. A.; Bray, R.C.; Diakun, G.P. *Biochem. J.* 1989, 260, 563-571
[63] Romão, M. J.; Archer, M.; Moura, I; Moura, J. J. G.; LeGall, Jean; Engh, R.; Schneider, M.; Hof, P.; Huber, R. *Science* 1995, 270, 1170-1176
[64] Huber, R.; Hof, P.; Duarte, R.O; Moura, J.J.G., Moura, I., Liu, M.Y., LeGall, J., Hille, R., Romão, M.J *Proc. Nat. Ac. Sci USA*,1996, 93, 8846-8851

[65] Jones, R. M.; Inscore, Frank E.; Hille, R.; Kirk, M. L. *Inorg. Chem.* 1999, 38, 4963-4970
[66] Ragsdale, S.W. *Critical Reviews in Biochemistry and Molecular Biology* 2004, 39, 165-195
[67] Gremer, D. H.; Kiefersauer, L.; Huber, R.; Meyer, O. *Proc. Nat. Ac. Sci. USA* 2002, 99, 15971-15976
[68] Gnida, M.; Ferner, R.; Gremer, L.; Meyer, O.; Meyer-Klaucke, W. *Biochemistry* 2003, 42, 222-230
[69] Bonin, I.; Martins, B. M.; Purvanov, V.; Fetzner, S.; Huber, R. Dobbek, H. *Structure* 2004, 12 1425-1435
[70] Mueller, A.; Diemann, E.; Rainer, J.; Boegge H. *Angew. Chem., Int. Ed.* 1981, 20, 934-955
[71] Wieghardt, K.; Hahn, M.; Weiss, J.; Swiridoff, W. *Z. Anorg. Allg. Chem* 1982, 492, 164-174
[72] Bristow, S.; Collison, D.; Garner, C.D.; Clegg, W. *Dalton Trans* 1983, 2495-2499
[73] Faller, J.W.; Ma, Y. *Organometallics* 1989, 8, 609-612
[74] Wilson, G.L.; Greenwood, R.J.; Pilbrow, J.R.; Spence, J.T.; Wedd, A.G. *J. Am. Chem. Soc.* 1991, 113, 6803-6807
[75] Laughlin, L. J.; Aston A. E.; Graham N. G.; Edward R. T. T.; Young, C. G. *Inorg. Chem.* 2007, 46, 939-948
[76] Doonan, C. J.; Nielsen, D. J; Smith, P. D.; White, J. W.; George, G. N; Young, C. G. *J. Am. Chem. Soc.* 2006, 128, 305-316
[77] Schindelin, H.; Kisker, C. ; Hilton, J.; Rajagopalan K. V.; Rees, D. C. *Science* 1996, 272, 1615-1621
[78] Czjzek, M.; Dos Santos, J.-P.; Pommier, J.; Giordano, G.; Méjean, V.; Haser, R. *J. Mol. Biol.* 1998, 284, 435-447
[79] Zhang, L.; Nelson, K.J.; Rajagopalan, K.V.; George, G.N. *Inorg. Chem.* 2008, 47, 1074-1078
[80] Dias, J.M.; Than, M. E.; Humm, A.; Huber, R.; Bourenkov, G. P.; Bartunik, H. D.; Bursakov, S.; Calvete, J.; Caldeira, J.; Carneiro, C.; Moura, J.J.; Moura, I.; Romão, M. J. *Structure* 1999, 7, 65-79
[81] Jormakka, M.; Richardson, D; Byrne, B.; Iwata, S. *Structure* 2004, 12, 95-104
[82] Ellis, P. J.; Conrads, T.; Hille, R.; Kuhn, P. *Structure* 2001, 9, 125-132
[83] Conrads, T.; Hemann, C.; George, G. N.; Pickering, I.J.; Prince, R. C.; Hille, R. *J. Am. Chem. Soc.* 2002 ,124 ,11276-11277
[84] Maher, M. J.; Santini, J.; Pickering, I. J.; Prince, R.C.; Macy, J. M.; George, G. N. *Inorg. Chem.* 2004, 43, 402-404

[85] Bruce, A.; Corbin, J.L.; Dahlstrom, P.L.; Hyde, R.; Minelli, M.; Stiefel, E. I.; Spence, J. T.; Zubieta, J. *Inorg. Chem.* 1982, 21, 917-926
[86] Liebeskind, L. S.; Sharpless, B. K.; Wilson, R. D.; Albers, J. *J. Am. Chem. Soc.* 1978, 100, 7061-7063
[87] Marabella, C. P.; Enemark, J. H.; Newton, W. E.; McDonald, J. W. *Inorg. Chem.* 1982, 21, 623-627
[88] Bristow, S.; Enemark, J. H.; Garner, C. D.; Minelli, M.; Morris, G. A.; Ortega, R. B. *Inorg. Chem.* 1985, 24, 4070-4077
[89] Mondal, J. U.; Schultz, F. A.; Brennan, T. D.; Scheidt, W. R. *Inorg. Chem.* 1988, 27, 3950-3956
[90] Jalal, L. S.; Mondal, U.; Uhrhammer, D.; Schultz, F. A. *Inorg. Chim. Acta* 1998, 278, 1-5
[91] Jalal, L. S.; Mondal, U.; Zamora, J. G.; Kinon, M. D.; Schultz, F. A. *Inorg. Chim. Acta* 2000, 309, 147-150
[92] Jalal; L. S.; Mondal, U.; Zamora, J. G.; Siew, S.-C.; Garcia, G.T.; George, E. R.; Kinon M. D.; Schultz, F. A. *Inorg. Chim. Acta* 2001, 321, 83-88
[93] Głowiak, T.; Jerzykiewicz, L.; Sobczak, J.M.; Ziółkowski, J. *Inorg. Chim. Acta* 2003, 356, 387-392
[94] Jalal, L. S.; Mondal, U.; Almaraz, E.; Bhat, N. G. *Inorg. Chem. Commun.*, 2004, 7, 1195-1197
[95] Hanna, T. A.; Ghosh, A. K.; Ibarra, C.; Zakharov, L. N.; Rheingold, A. L.; Watson, W. H. *Inorg. Chem.* 2004, 43, 7567-7569
[96] Boyd, I. W.; Dance, I. G.; Landers, A. E.; Wedd, A. G. *Inorg. Chem.* 1979, 18, 1875-1885
[97] McNaughton, R. L.; Tipton, A. A.; Rubie, N. D.; Conry, R. R.; Kirk, M. L. *Inorg. Chem.* 2000, 39, 5697-5706
[98] McMaster, J.; Carducci, M. D.; Yang, Y.-S.; Solomon, E. I.; Enemark, J.H. *Inorg. Chem.* 2001, 40, 687-702
[99] Bradbury, J. R.; Wedd, A. G.; Bond, A. M. *Chem. Commun.* 1979, 1022-1025
[100] Wang, X.-B.; Inscore, F.E.; Yang, X.; Cooney, J.J.A.; Enemark, J.H.; Wang, L.-S. *J. Am. Chem. Soc.* 2002, 124, 10182-10191
[101] Davie, S.R.; Rubie, N.D.; Hammes, B. S.; Carrano, C. J.; Kirk, M. L., Basu, P. *Inorg. Chem.* 2001, 40, 2632-2633
[102] Nemykin, V. N.; Davie, S. R.; Mondal, S., Rubie, N.; Kirk, M. L.; Somogyi, A.; Basu, P. *J. Am. Chem. Soc.* 2002, 124, 756-757
[103] Boyde, S.; Ellis, S. R.; Garner, C. D.; Clegg, W. *Chem. Commun.* 1986, 1541-1543

[104] Davies, E. S.; Beddoes, R. L.; Collison, D.; Dinsmore, A.; Docrat, A.; Joule, J. A.; Wilson, C. R.; Garner, C. D. *Dalton Trans.* 1997, 3985-3996
[105] Oku, H.; Ueyama, N.; Kondo, M.; Nakamura, A., *Inorg. Chem.* 1994, 33, 209-216
[106] Ansari, M. A.; Chandrasekaran, J.; Sarkar, S. *Inorg. Chim. Acta* 1987, 133, 133-136
[107] Coucouvanis, D.; Hadjikyriacou, A.; Toupadakis, A.; Koo, S.-M.; Ileperuma, O.; Draganjac, M.; Salifoglou, A. *Inorg. Chem.* 1991, 30, 754-767
[108] Oku, H.; Ueyama, N.; Nakamura, A. *Inorg. Chem.* 1997, 36, 1504-1516
[109] Matsubayashi, G.; Nojo, T.; Tanaka, T. *Inorg. Chim.* Acta 1988, 54, 133-135
[110] McCleverty, J. A.; Locke, J.; Ratcliff, B.; Wharton, E. J. *Inorg. Chim. Acta* 1969, 3, 283-286
[111] Das, S. K.; Chaudhury, P. K.; Biswas, D.; Sarkar, S. *J. Am. Chem. Soc.* 1994, 116, 9061-9070
[112] Donahue, J. P.; Goldsmith, C. R.; Nadiminti, U.; Holm, R. H. *J. Am. Chem. Soc.* 1998, 120, 12869-12881
[113] Sugimoto, H.; Harihara, M.; Shiro, M.; Sugimoto, K.; Tanaka, K.; Miyake, H.; Tsukube, H. *Inorg. Chem.* 2005, 44, 6386-6392
[114] Lim, B. S.; Donahue, J. P. Holm, R. H. *Inorg. Chem.* 2000, 39, 263-273
[115] Lim, B. S.; Holm, R. H. *J. Am. Chem. Soc.* 2001, 123, 1920-1930
[116] Wang, J. J.; Tessier, C.; Holm, R. H. *Inorg. Chem.* 2006, 45, 2979-2988
[117] Johnson, M. K.; Rees, D. C.; Adams, M. W. W. *Chem. Rev.* 1996, 96, 2817-2839
[118] George, G. N.; Prince, R. C.; Mukund, S.; Adams, M. W. W. *J. Am. Chem. Soc.* 1992, 114, 3521-3523
[119] Hu, Y.; Faham, S.; Roy, R.; Adams, M. W. W.; Rees, D. C. *J. Mol. Biol.* 1999, 286, 899-914
[120] Raaijmakers, H.; Macieira, S.; Dias, J.M.; Teixeira, S.; Bursakov, S.; Huber, R.; Moura, J. J.G; Moura, I.; Romão, M. J. *Structure* 2002, 10, 1261-1272
[121] George, G.N.; Colangelo, C. M.; Dong, J.; Scott, R. A.; Khangulov, S. V.; Gladyshev V. N.; Stadtman T. C. *J. Am. Chem. Soc.* 1998, 120, 1267-1273
[122] Boyington J. C.; Gladyshev V. N.; Khangulov S. V.; Stadtman T. C.; Sun P. D. *Science* 1997, 275, 1305-1308
[123] Raaijmakers, H. C. A.; Romão, M. J. *J. Inorg. Biochem.* 2006, 11, 849-854
[124] Barber, M. J. *Biochemistry* 1986, 25, 150-158
[125] Chen, G.J.-J.; McDonald, J. W.; Newton, W. E. *Inorg. Chim. Acta* 1976, 9, L67-L68

References

[126] Cervilla, A.; Llopis, E.; Ribera, A.; Domenech, A.; Sinn, E. *Dalton Trans.* 1994, 3511-3516
[127] Yu, S.-B.; Holm, R.H. *Inorg. Chem.* 1989, 28, 4385-4391.
[128] Ueyama, N.; Oku, H.; Nakamura, A. *J. Chem. Am. Soc.* 1992,114, 7310-7311
[129] Das, S.K.; Biswas, D.; Maiti R.; Sarkar, S. *J. Chem. Am. Soc.* 1996, 118, 1387-1397
[130] Sarkar, S.; Das, S.K. *Proc. Ind. Ac. Sci. – Chem. Sci.* 1992,104, 533-534
[131] Lim, B. S.; Sung, K.-M.; Holm, R. H. *J. Am. Chem. Soc.* 2000, 122, 7410-7411
[132] Sugimoto, H.; Sugimoto, K. *Inorg. Chem. Commun.* 2008, 11, 77-80
[133] Groysman, S.; Holm, R. H. *Inorg. Chem.* 2007; 46, 4090-4102
[134] Holm, R.H. *Coord. Chem. Rev.* 1990, 100, 183-221
[135] Schultz, B. E.; Holm, R. H. *Inorg. Chem.* 1993, 32, 4244-4248
[136] Schultz, B. E.; Hille, R.; Holm, R. H. *J. Am.* Chem. Soc. 1995,117,827-828
[137] Topich, J.; Lyon, J.T. *Inorg. Chem.* 1984, 23, 3202-3206
[138] Dahlstrom, P. L.; Hyde, J. R.; Vella, P. A.; Zubieta, J. *Inorg. Chem.* 1982, 21, 927-932
[139] Craig, J.A.; Harlan, E. W.; Snyder, B. S.; Whitener, M. A.; Holm, R. H. *Inorg. Chem.* 1989, 28, 2082-2091
[140] Ueyama, N.; Oku, H.; Kondo, M; Okamura, T.; Yoshinaga, N.; Nakamura, A. *Inorg. Chem.* 1996, 35, 643-650
[141] Oku, H.; Ueyama, N.; Nakamura, A. *Inorg. Chem.* 1995, 34, 3667-3676
[142] Najmudin, S.; Gonzalez, P. J.; Trincao, J.; Coelho, C.; Mukhopadhay, A.; Cerqueira, N. M. F. S. A.; Romão, C. C., Moura, I.; Moura, J. J. G.; Brondino, C. D.; Romão, M. J. *J. Biol. Inorg. Chem.* 2008, 13, 737-753
[143] Romão, M. J. *Dalton Trans.* 2009, 4053-4068
[144] Ma, X.; Schulzke, C.; Yang, Z.; Ringe, A.; Magull, J. *Polyhedron* 2007, 26 5497-5505
[145] Ma, X.; Schulzke, C.; Schmidt, H.-G.; Noltemeyer, M. *Dalton Trans.* 2007, 1773-1780
[146] Sugimoto, H.; Tarumizu, M.; Miyake, H.; Tsukube, H. *Eur.J. Inorg. Chem.* 2006, 4494-4497
[147] Sung, K.-M.; Holm, R. H. *Inorg. Chem.* 2001, 40, 4518-4525
[148] Sung, K.-M.; Holm, R. H. *J. Am. Chem. Soc.* 2002, 124, 4312-4320
[149] Sung, K.-M.; Holm, R. H. *J. Am. Chem. Soc.* 2001, 123, 1931-1943
[150] Collison, D.; Garner, C. D.; Joule, J. A. *Chem. Soc. Rev.* 1996, 25, 25-32
[151] Bertero, M. G.; Rothery,R. A.; Palak, M.; Hou, C.; Lim, D.; Blasco, F.; Weiner, J. H.; Strynadka,N. C. J. *Nature Struct. Biol.* 2003, 10, 681-687

[152] Bradshaw, B.; Dinsmore, A.; Garner C. D.; Joule, J. A. *Chem.Commun.* 1998, 417-418;

[153] Bradshaw, B.; Collison, D.; Garner C. D.; Joule, J. A. *Chem. Commun.* 2001, 123-124

[154] Marbella, L.; Mitasev, B. S.; Basu, P. *Angew. Chem. Int. Ed.* 2009, 121, 4056-4058

INDEX

A

absorption, 26, 33, 39, 52
absorption spectroscopy, 25, 33, 39, 46, 52
acceptor, 16, 52
acceptors, 53
acetaldehyde, 20, 47
acetonitrile, 31, 53, 55
acetylene, 19, 20, 47
acid, 13, 20, 26, 38, 39
activation, 34
active site, xi, 13, 16, 17, 18, 19, 20, 21, 24, 26, 27, 29, 30, 33, 37, 39, 41, 42, 44, 48, 49, 51, 54, 55, 59, 62
Adams, 65, 67, 70
adenine, 33
adenosine, 18
aerobic, 34
agent, 34, 41, 57
agents, 53
alanine, 38
alcohol, 17, 59
aldehydes, 45
ambiguity, 26, 45
amino, 13, 19, 20, 26, 38, 49, 54
amino acid, 13, 19, 20, 26, 38, 49, 54
amino acids, 19, 49, 54
anaerobic, 15, 52
anomalous, 37

aqueous solution, 61
archaea, 15
arsenite, 20, 38, 44, 55
aspartate, 20, 38
assignment, 34
assumptions, 17
atmosphere, 15
atoms, 17, 25, 28, 34, 37, 38, 42, 46, 47, 54
availability, 15, 16, 24

B

back, 38, 55, 57
bacteria, 16
behavior, 16, 29, 42
benzene, 30
benzoquinone, 57, 63
binding, 37, 54
biological systems, 16
bonding, 23, 40
bonds, 20, 37, 38
borate, 28, 35, 56, 64
Boron, v
buffer, 45

C

carbon, iv, 13, 17, 59
carbonyl groups, 43

carboxylic, 45
carboxylic acids, 45
catalysis, 21, 26, 57
catalyst, 29, 52
catalytic properties, 28
cell, 13
chemical oxidation, 55
chicken, 17, 26
chloride, 25
circular dichroism, 34
cis, 29, 35, 39, 40, 41, 42, 53
classes, 19
clusters, 20
CO_2, iii, 46, 48
cofactors, 13, 20, 21, 24, 55
Columbia, 23
community, 20
composition, 33, 37, 44, 45, 59
compounds, xi, 13, 20, 21, 28, 30, 39, 41, 42, 43, 44, 47, 48, 54
comprehension, 21
concentration, 15, 25, 47, 61
configuration, 53
Congress, viii
conjugation, 59
control, 40, 53
conversion, 46, 47, 55
copper, 34
couples, 42, 51, 53, 57
coupling, 34, 59
covalent, 38
creativity, 62
crust, 15
crystal structure, 17, 18, 19, 26, 34, 37, 38, 40, 42, 46, 47, 48, 49, 59, 62
crystal structures, 19, 34, 40, 42, 59
crystallization, 42, 47
crystals, 34
curiosity, 24
cyanide, 46
cysteine, 20, 26, 38, 47, 49, 54, 55, 63
cytochrome, 25, 27
cytosine, 18, 33

D

decane, 52, 64
definition, 34
dehydrogenase, 19, 20, 34, 46, 48, 63
delocalization, 59
density, 34, 45
derivatives, 17, 34
dimer, 53
dimeric, 53, 55
dimerization, 26, 28, 35, 53
dinucleotides, 37
dissociation, 53
DMF, 29
donor, 16, 43
donors, 26, 30, 55

E

E. coli, 46, 63
earth, 15
electron, 13, 16, 17, 23, 25, 27, 28, 29, 38, 40, 45, 51, 56, 59, 64
electron density, 45
electronic structure, 23, 40
employment, 21
enthusiasm, 62
environment, 13, 15, 20, 21, 26, 30, 38, 49
enzymatic, 13, 20, 40, 42, 44, 52, 55, 62
enzymes, xi, 13, 15, 16, 17, 18, 19, 20, 21, 25, 26, 28, 29, 30, 33, 34, 37, 38, 39, 40, 42, 44, 45, 46, 47, 49, 51, 54, 56, 59, 62
EPR, 28, 34
Escherichia coli, 38, 46, 63
ethylenediamine, 29
eukaryotes, 16, 18
evolution, 15
EXAFS, 20, 25, 37, 44, 45, 63
expertise, 62

F

family, 18, 19, 25, 26, 28, 30, 33, 34, 37, 38, 39, 40, 42, 44, 45, 46, 47, 49, 51, 54, 55, 56, 62
family structure, 33
fine tuning, 57, 59
fluorescence, 17
formaldehyde, 46

G

glycerol, 45
greenhouse, 46
greenhouse gas, 46
groups, 13, 23, 26, 28, 33, 37, 43, 45, 47, 51, 62
guanine, 33, 37, 63

H

habitat, 13
handling, 60
Harvard, 21, 23
heme, 20
heterogeneous, 46
histidine, 17
human, 13, 15, 26
hybrid, 24
hydration, 20
hydride, 34
hydroxide, 20, 31, 44
hydroxyl, 46

I

imagination, 62
inactivation, 46
inactive, 33, 55
inclusion, 43
incubation, 46
injury, viii
innocence, 23

inorganic, 21, 34, 42, 47, 62
insertion, 39, 64
insertion sequence, 39, 64
insight, 51
instability, 20
interaction, 17, 35
interactions, 23
Investigations, 17
ions, 15
iron, 17
isoenzymes, 47
isomerization, 54
isomers, 40, 42
isotope, 29, 52, 55

J

justification, 21

K

kinetics, 55

L

labeling, 55
ligand, 13, 17, 20, 21, 23, 26, 27, 28, 30, 33, 34, 37, 38, 40, 42, 43, 44, 46, 48, 49, 53, 54, 57, 59, 60, 61
ligands, 13, 18, 20, 23, 25, 28, 30, 37, 38, 42, 43, 45, 47, 48, 49, 53, 54, 57, 60
limitations, 20, 37
linkage, 38
liver, 17, 26
lying, 26, 30, 59

M

magnesium, 45
magnetic, viii, 23, 34
mass spectrometry, 17, 52
media, 55
metabolism, 13

metals, 15, 20, 23, 61
methane, 40
methanol, 28
microorganisms, 13, 15
milk, 34
mimicking, 20, 31, 48
model system, 20, 21, 29, 56
modeling, 26, 39, 42, 43
models, 13, 20, 42, 47, 48, 55
molecules, 13, 16, 37, 46
molybdenum, xi, 13, 15, 16, 17, 18, 19, 20, 21, 24, 25, 27, 28, 29, 30, 33, 35, 37, 38, 40, 42, 44, 45, 46, 47, 49, 51, 54, 57, 59, 60, 62
mutant, 26

N

NADH, 25
naphthalene, 40
natural, 20, 28, 47, 51, 59
New Jersey, 66
New York, vii, viii
Newton, 69, 70
Nielsen, 68
NIR, 39
nitrate, 19, 25, 38, 54, 59
nitrogen, 13, 16
nitrogen fixing, 16
nucleotides, 18
nucleus, 17, 59

O

observations, 46
organ, 34
organic, 17, 61, 62
organometallic, iii, 34
orientation, 30, 41, 42
oxidation, 13, 15, 16, 20, 23, 25, 29, 35, 38, 40, 42, 46, 47, 48, 55, 56, 60
oxidative, 15, 25
oxide, 37, 48, 57, 64
oxides, 44

oxo groups, 26, 28, 33, 37
oxygen, 13, 15, 16, 25, 28, 29, 35, 37, 40, 41, 43, 46, 49, 51, 52, 55, 56, 57
Oxygen, 29, 51, 63

P

pathways, 30
peptide, 39, 44
phenol, 56
PhOH, 43, 49
phosphate, 17, 18, 45
photosynthetic, 15
physiological, 25, 38
planar, 23, 59
play, 41, 44
polypeptide, 37, 38
poor, 42
prevention, 53
probe, 40
prokaryotes, 16, 18
prokaryotic, 33, 37
protein, 20, 26, 38, 45
protein crystallography, 20
protons, 60
Pseudomonas, 34
PTA, 52, 53, 56, 64
purification, 38
pyramidal, 26, 33, 38, 42, 44

R

Raman, 26
range, 61
reactants, 51
reaction mechanism, 52
reaction rate, 54
reactivity, 21, 23, 26, 49, 54
recovery, 17
redox, 16, 17, 23, 26, 38, 40, 42, 46, 48, 51, 53, 55, 56, 57, 59
reductases, 25, 54
regeneration, 28, 29, 41, 52
relationships, 21, 42

relatives, 47
relevance, 28, 30, 52, 55
residues, 20
resolution, 39, 46
respiratory, 38
rings, 35, 59, 60
room temperature, 28

S

scarcity, 16
Schiff, 40, 48
Schiff base, 40, 48
Schmid, 65
selenium, 39, 43, 46, 54
series, 40, 48
serine, 20, 26, 37, 38, 40
services, viii
signals, 34, 40
similarity, 30, 47, 49
sites, 21, 38, 45, 49, 59, 62
solubility, 15
solvent, 29
solvents, 29
species, 15, 26, 29, 34, 35, 37, 38, 41, 43, 48, 49, 53, 55, 56, 57
specificity, 20
spectroscopy, 26
spectrum, 52
stages, 21, 23
steric, 28, 44, 53
stoichiometry, 24
storage, 45
stress, 39
students, xi
substitution, 30
substitution reaction, 30
substrates, 17, 37, 51, 52, 57
sulfate, 16, 25
sulfur, 13, 15, 17, 25, 28, 34, 35, 37, 38, 42, 46, 47, 49, 54
sulphur, 38
Sun, 70
superposition, 38
synthesis, 44, 48, 49, 54, 66

T

temperature, 16, 28
toluene, 29
trans, 29, 34, 40, 41, 42, 53
transfer, 13, 16, 21, 25, 26, 27, 28, 29, 34, 35, 40, 41, 47, 48, 49, 51, 52, 54, 55, 56, 57, 62, 63, 64
transformation, 25, 53
transition, 13, 15, 23
transition elements, 15
transition metal, 23
transitions, 20
trial, 20
trial and error, 20
trimethylamine, 37
tungsten, xi, 13, 15, 16, 17, 18, 19, 20, 21, 24, 43, 45, 46, 47, 48, 49, 51, 55, 57, 59, 60, 62
turnover, 28, 34, 38, 41, 52, 60

U

uncertainty, 20

V

variability, 21
versatility, 16, 20

W

water, 13, 15, 16, 20, 26, 28, 29, 38, 41, 47, 55, 57
weak interaction, 35
weathering, 15
wild type, 26
workers, 13, 17, 24, 34, 35, 40, 46, 48, 52, 61

X

X-ray absorption, 25, 38, 46, 63
X-ray crystallography, 41
X-ray diffraction, 45

Y

yield, 44, 48, 57, 60